DARWIN'S FLAW

Man's Superiority to Women in Mate Selection

William Spriggs

NEW WAVE
Books

DARWIN'S FLAW
MAN'S SUPERIORITY TO WOMEN IN MATE SELECTION
BY WILLIAM SPRIGGS

For more information, please contact:
Darwin's Flaw/ New Wave Books
P.O. Box 147149
Lakewood, Colorado 80214

web: www.darwinsflaw.com

To males who are frightened by this coming social upheaval, Mr. Spriggs tells them—it won't happen for another 200 to 250 years.

He advises females who clamor for equality and a reversal of gender discrimination now to be patient. On the Big Bang Cosmic Calendar Clock that measures time since the Big Bang to the present—this will occur in a mere half second.

In 1615, Galileo confirmed Copernicus's theory that the Earth revolved around the Sun.

This was the exact opposite of the overwhelming cultural belief that Earth, created by God, was the center of the universe, and that the Sun revolved around the Earth as its "auxiliary."

It stunned the dominant established religious institution so much that Galileo was placed under house arrest, and his findings were kept secret for about 250 years.

Along with the established view that Earth was the center of the universe, there was also a deep cultural belief that the male of our species was the masculine representative of the male God on this Earth—male-kind—and he had dominance over all the creatures large and small on this planet including the female of our species.

The established belief was that the female of our species was merely a receptacle of male-kind's "superior" seed, and that she revolved around the male at the center of his world as his "auxiliary."

Darwin's Flaw takes a similar anti-establishment tone to Galileo's, in that it argues that the established cultural belief of the female as the helper of the male, revolving around the male as the "center" of all things, is wrong, and that actually the male followed the female out of the jungle to create our human species—but most importantly: that the male is destined to revolve around the female as an auxiliary to assist her in the future.

William Spriggs takes the reader on a voyage between those two time-measured events and argues how it began, how our species adapted to changes, and what the future holds.

Dedicated to my wife, Diana
For her support and encouragement over the many years
that this project took

CONTENTS

ACKNOWLEDGEMENTS

This book would not have been at all possible without acknowledging my partner, companion, lover, and understanding friend—my wife, Diana. Her patience in allowing me to sit in front of this device called a word processor and screen which is hooked up to something called the Internet that allowed me—a simple, blue-collar man of toil—who depended on his shoulders and back for a livelihood—the opportunity to make *Darwin's Flaw* a physical reality. Her encouraging comments throughout the years while I planned, sketched out, wrote, and put together this book have been invaluable.

Diana has been the inspiration for the book and the glue that kept it all together. It was by observing her and understanding her determination through difficult years in our lives revolving around her daughter and grandson that lit the imaginary lightbulb that illuminated the wide-reaching implications of the patient accomplishments behind the female of our species.

I suppose one could view this book as a new trend in book publications: aiding an unknown and naive author assistance through the myriad obstacles of attempting to reach a larger, targeted and receptive audience without taking the traditional route of seeking an agent and hoping for positive results. As such, I cannot praise the good folks at Good Book Developers in Atlanta, Georgia whose motto is a simple: We Make Good Books; they've have made that part of the voyage very easy.

They proofread my work, designed the book cover, they acquired ISBN numbers, placed my book for sale on CreateSpace.com and walked me through the process, keeping me fully informed at each step. They did all the hard stuff that authors never think about (and why should they?). I think of them as my personal publishing concierge.

Outside my personal realm, I want to think all those people in the evolutionary psychology movement who understand and had the courage to posit as a bedrock principle that our current human behavior is anchored biologically to our primate past.

I wish to thank the U.S. Postal Service for granting me the privilege of 21 years of steady employment and an adequate pension to continue my work without the adversity of struggling for existence to work in relative comfort. I also need to thank the organization for allowing me to meet my future wife and all the friends I have made inside this highly-reputable and long-standing institution.

But most importantly, since 1986 when I started in the Post Office until I retired in 2007, this occupation allowed me the honor of working side-by-side with women who earned the same salary as myself—to the penny—in fact, my wife, out-earned me for many years because she also put in more overtime hours than yours truly.

And how could I not acknowledge the American Defense Department for the development of the Internet, through which I could search worldwide and find the materials I needed to make this book possible. An additional tip of the hat to Google, for giving me, a poor person with limited resources, a powerful search engine allowing me to gather all of this amazing material.

INTRODUCTION

WAITING FOR THE NEXT WAVE

At one time in history, humanity believed that the earth was the center of all celestial bodies. This model served as the predominant cosmological system in many ancient civilizations.

In 1615, Galileo confirmed Copernicus's theory that the earth did indeed revolve around the sun. The Catholic Church did not take too kindly to Galileo's confirmation of the reversal of embedded cultural wisdom that threatened their dominance. They promptly placed Galileo on trial, branded him a heretic, and placed him under home arrest until his death. The Catholic Church did not reverse its position for over 250 years.

It was a difficult concept to reverse because if you were a simple man of the earth, plowing your field, you could easily observe that the sun rose in the east and set in the west while you remained relatively in one place.

Your religious leaders confirmed this observation as proof of God's wisdom and power; refusal to believe this fact meant you were in danger of being excommunicated, which led to your immortal soul being sent to hell.

I believe that we are at a similar moment in history, when another wide-spread belief embedded in our culture is about to come under close scrutiny, and, as I will attempt to prove, it may very well be at the early stages of reversal.

I will strongly argue that it was the female who led us out of the jungle, and that the long established belief by many scientists (including Darwin) that the male is more intelligent than the female is seriously flawed.

Perhaps some will argue I unfairly picked Darwin as the pivot point on this subject, and I have not done so because of his thoughts about evolution, but primarily because of his time and place in history when England and males were at their pinnacle of social, political, economic, industrial, and military influence. This book is also about what happened to the co-creator of the natural selection theory and why that person is an important social footnote in history.

Because lurking in the background of *Darwin's Flaw* about male choice in mate selection, there is a "bigger picture" of human exclusionary behavior deep within our biological make-up. After the exit from the jungle through cooperative hunter-gatherer behaviors, this deep exclusionary behavior reemerged with great influence when we stopped migrating from location to location, began to live in groups, and created surplus resources through the use of the heavy plow—a device used primarily by the male. From this point in history, with the female still adjusting to her secondary role (the 2nd sex), the male seized the mantle as the superior gender.

Today this exclusionary behavior has evolved into social discriminations of multiple layers that hide in plain sight. To unravel these behaviors entails us diving into today's science of sociology as we look back at the cultural and social climate of Darwin's Victorian England, in light of which, I argue, Darwin's hierarchical position and gender had a major influence on his scientific conclusions regarding the Victorian male's "choice" of a mating partner.

Darwin's Flaw is a tale that needs to be told. It is time to become social carpenters and restore a more balanced view of our hierarchal discriminations against all people we consider "below us." But the primary message you will take away from *Darwin's Flaw* is that we will be placing greater importance on the female's role in our evolutionary voyage as we transition into the future.

For the female shall be the future evolutionary leader of our species—once again.

CHAPTER 1

DARWIN'S FLAW

Man's superiority to women in mate selection

*"… Darwin's theories of variation needed sexuality.
Without sexual reproduction he had no mechanism for
the spontaneous appearance of variations in the offspring;
without variations there was no possibility of selection or
of transmutation. Sexual relations between plants, and
between animals, and between humans, were the essential footing to his version of evolution."*
(Browne, 1995 p. 528)

As you gathered from the opening quote from Janet
Browne's masterful two-volume biography of Charles Darwin, this book has a central meme: how we traversed the
path of life from our primate origins up to our present
moment in time, in order to reproduce and pass on our
genetic heritage. The primary method our species relies
upon and prefers, despite all the advances in science and
social changes, is still the old-fashioned method: via sex of
the two genders.

This book is about the meaning of life surrounding
our human ancestral beginnings and how it evolved from
our jungle origins. It continues with the domestication of

plants and animals and most prominently looks closely at a socially constructed world created by our species that affects every person at every social level of our modern societies. It also dares to speculate on the future of our species: led by females—again.

But I want you to remember an important fact that lies beneath the surface of this book: yes, it is about producing progeny, but since we are an advanced species, it's also about protecting them and giving them the ability not just to survive but to prosper.

We have much to cover, so let's jump right into the flaw surrounding Darwin's male as the selector of mates and the other various reasons for this book.

In the simplest of explanations, Darwin flipped human evolutionary behavior on its head when he wrote in his book *The Descent of Man and Selection in Relation to Sex* in 1870 that the modern human male won the exclusive right to choose his female mating partner because he had a higher intellect. This higher intellect, he reasoned, was based on the learned experiences of hunting and the skills that were passed on by his ancestors throughout the millennia.

This is in direct opposition to his own declaration in his 1859 book, *On the Origins of Species by Means of Natural Selection, or the Preservation of the Favored Races in the Struggle for Life*, that in the animal world it was the female that was the primary selector of the mate.

Something happened in the transition, and I posit that it was not science but culturally and socially constructed bias of the male gender's long history of dominance over the female—especially in Darwin's moment in history.

But I must add a caveat: it happened because the female wanted this to occur—temporarily. When I use the word "temporarily," I am using time as measured on Carl Sagan's, the Big Bang Cosmic Calendar Time Clock (Sagan, 1980) which starts at the Big Bang some 13.8 billion years ago and reduces time down to an analog clock beginning on December 31st, where one second on the Cosmic Clock equals 500 years in our time. Keep that thought in mind, as we will return to this time meme later. A meme is a cultural idea or thought that is transmitted and replicated similar to the transmission of genes.

Darwin's thought process was that the males' intellect was sharpened by stalking prey, strategic planning, and cooperative male bonding through the creation of language used during hunting. His reasoning continued with the premise that with this advanced intellect the human male created all the wonders of the modern industrial world that everyone saw around them. Darwin observed, and then believed, that the modern, educated Victorian male was atop the evolutionary tree, and that he, and others males like him, were there because it was "the natural order of things."

I'm not the only one who noticed this flaw. Dr. Adrienne Zihlman, when writing about the same subject in *Woman the Gatherer* (Dahlber, 1981), noted that there was a "curious discrepancy" about the theory. Recently, Dr. Philippa Hardman, Research Associate of the Darwin Correspondence Project at Cambridge University in England, produced a short film entitled *Darwin's Women,* in which she noted from the book *Descent:* "...*which was explicit in its content of sexual display and arguing that sexual selection was almost exclusively driven by female aesthetic tastes—or at least in every species except humans. Interestingly, Darwin didn't apply the same idea to his own species.*"

Ms. Hardman also noted in the film that Darwin publicly attacked John Stuart Mills on his treatise "Equity for Women," but then teaches us that Darwin wrote to over

150 female scientists and encouraged them in their endeavors while at the same time still *"toed the conservative line"* in public when it came to the Victorian female's secondary place in society. These are concrete examples, which I hold faith in, of Darwin molding his thought processes to conform to the social norms of his social hierarchical ranking in his timeline. (Cambridge University, 2013).

* * *

Thomas Huxley, who was later to be called Darwin's Bulldog because of his tenacious defense and support of Darwin's natural selection theory after it was published, wrote an essay entitled "On the reception of the *Origin of Species*," and therein wrote of himself: "How incredibly stupid not to have thought of that" (Huxley, 1887).

That sentence poked at me from across the ages and stirred me to explore this in more detail. I wanted to know why more scientists have missed this glaring flaw. Were they all being incredibly stupid? How could so many highly educated males across the ages overlook this and not help reverse the many oppressive subjugations against women, who are the primary gender responsible for nurturance of our species?

I believe female choice is a quintessential principle set in evolutionary stone. Although we will learn later through new scientific studies that males do have a substantial part in the mate selection process, it is still the female who must make the more difficult decision regarding mate choice because she has the more laborious task ahead of devoting time and energy to her future progeny. I don't know if every female is wholly conscious of this, but I believe the decision she makes transcends not just her own progeny but the future of our own species as a whole. I strongly believe it was the female of our species—through her cooperative and nurturing nature, and yes, superior intelligence—who was responsible for our split from the primates and began our journey on the path to humanhood. More on that in a short while.

If you believe this is something you can seriously wrap your head around, you can now proudly call yourself an evolutionary feminist—or perhaps, an evolutionary humanist. Feminism has many layers and explanations and has had different objectives throughout its short time on the cosmic calendar clock, but this book will never lose sight of the primary idea that it is the female that gave the push to our species in the direction of becoming human.

THE PEACOCK MADE HIM DO IT

I think it is vitally important that we crystalize the manner in which Darwin came to the idea of sexual selection of the female in the animal world in the first place. This is an important moment for our human species because it allowed Darwin to peek into the doorway of the dawning of the human mind as it began to separate from the "nature" side of the animal mind to the "nurture" side of the human mind.

It was an agonizing moment of clarity for Darwin when he eliminated all possible conclusions regarding natural selection and forced himself to create a new theory that was separate from the "survival of the fittest" theory in the animal world. Darwin took a new path that digressed from the physical. This new theory had to do with choice: Sexual Selection.

> "Man is more powerful in body and mind than woman, and in the savage state he keeps her in a far more abject stage of bondage than does the male of

any other animal; therefore it is not surprising that he should have gained the power of selection" (Darwin, 1871, p. 901).

"To avoid enemies, or to attack them with success, to capture wild animals, and to invent and fashion weapons, requires the aid of the higher mental faculties, namely, observation, reason, invention, or imagination. These various faculties will thus have been continually put to the test, and selected during manhood...the characters thus gained will have been transmitted more fully to the male than to the female offspring...thus man has ultimately become superior to woman" (Darwin, 1871, p. 400).

* * *

To further shed light on his deduction, let's quote from Anne Browne's masterful two-volume biography, *Charles Darwin, The Power of Place*:

"The possibility of female choice among humans hardly ruffled the surface of his argument, although he repeatedly claimed that female choice was the primary motor for sexual selection in animals. Primitive societies, he conceded, may be matriarchal or polygamous. However, he regarded this as an unsophisticated state of affairs, barely one step removed from animals. Advanced human society, to Darwin's mind, was patriarchal, based on what was then assumed about primate behavior and the so-called 'natural' structure of civilized societies. For Darwin, it was self-evident that in civilized regimes men did the choosing. But his vision of mating behavior was an explicit expression of his class and gender...For him, Victorian males set the evolutionary compass" (Browne, 2002, pg. 346).

It is important to note how influential this particular male-dominated hunter theory became as noted by Sarah Blaffer Hrdy:

"Thus did the 'hunting hypothesis' morph into one of the most long-lasting and influential models of behavior in anthropology. Subsequent versions wove together increasingly coherent scenarios in which early human evolution was a 'direct consequence of brain expansion and material culture' fueled by an increasingly bipedal, increasingly effective hunter. Big brains, and with them superior intelligence, were viewed as 'the sine qua non of human origin'" (Hrdy, pg. 147, 2009).

Even though the theory about males' superior intellect gained through hunting skills persisted for over one hundred years, it seemed to have reached a fevered pitch during the 1960s in America, in what one could only call the "Man the Hunter" era.

In this American time period, the country was still atop the economic world, as much of Germany and Japan were still in their recovery stage from WWII and corporate America had not yet discovered cheap labor in China. America was in love with cheap gas and "muscle cars"; it was undergoing social upheaval in the wake of the prolonged Vietnam War with war images and body counts drifting across TV screens every night. On top of that, with the arrival of the civil rights movement came images across the same screens of white males showing us their worst in keeping "people down."

Not much changed until a new crop of scientists trained to observe subsistent strategies of modern hunter-gatherers became very skeptical of the Man the Hunter theory and for many years chaffed over this masculine superiority through hunting, thinking it had gone on long enough. It wasn't until around the late 1990s when behavioral ecologist/archaeologists began to gather courage and present their evidence. It continued with heated exchanges within the evolutionary community, and somewhere around 1999, the theory had collapsed. I suspect and hope that more will be written in the future about these events that occurred below the surface, which will benefit all of us who seek the truth.

Again, Professor Hrdy:

> "Like all young primates, *Homo ergaster* [that is, the African branch of *Homo erectus*] juveniles probably had to eat several times a day, every day. Like modern children, they probably relied on others to provide most of their food for years after weaning. The hunting hypothesis holds that early human males were the main source of this support, yet traditional East African hunters living in similar habitats today cannot meet this need, despite their use of sophisticated weapons. Though meat represents a sizable fraction of their families' annual caloric intake, it is not reliably enough to satisfy the daily nutritional needs of their children" (Hrdy, 2009, pg., 149).

So, if "Man the Hunter" did not fulfill the daily nutritional needs of the new human child, then we are left with the obvious "undiscovered" explanation. Yes, my friends, it was the daily gathering of herbs, nuts, fruit, and roots found by the patient and persistent female who saved the species from starvation. Somehow male anthropologists since

Darwin's pronouncement of the hunting male's superiority "missed" that small detail in their studies. I assume that they were too busy studying the males as they prepared for and witnessed the strategies behind the actual hunt while ignoring the females remaining in camp and providing for the children's daily needs. I have to stop here and remind you that this discovery of the flaw in Darwin's theory has been handed down to us just shy of 50 years ago.

So if Darwin was mistaken because he was overwhelmed by his social environment, why has the modern female not come forward and denounced this flip in sexual mate selection and stormed the barricades hiding truth and reason?

* * *

We will study more closely this class and gender matter surrounding the Victorian era later, but can't you see how Darwin could come to no other conclusion in his Victorian worldview? Didn't males run and control the planet throughout recorded history? Didn't males invent the marvelous machines that powered the new Industrial Age? Weren't males in authority in the government, military, and local judiciaries? Didn't they create all the awe-inspiring art treasures, poetry and symphonies that were handed down through the ages? Didn't they plan and build all the surrounding infrastructures of roads, bridges, and aqueducts? Hadn't they guided us through the ages with religious values and morality? And weren't Buddha, Mohammed, Jesus, and God all male?

The list could go on and on, leading males to the inevitable conclusive question: what flaw?

* * *

Before we dive deeper into Darwin's flaw, we must look briefly at the other theory that preceded Darwin's Sexual Selection Theory: the Natural Selection Theory.

We all know that Darwin's Natural Selection Theory, published in 1859 as *On the Origin of Species by Means of Natural Selection, or the Preservation of Favoured Races in the Struggle for Life*, has held up nicely after 157 years in the scientific community despite the mud-slinging insults and denials brought on by religious fundamentalists. You must always remember this meme—and I will repeat it several times throughout the book: a good gig with benefits is hard to give up without a fight.

Natural Selection is basically the random series of events in one's local environment that causes the species and their succeeding progeny to adapt physically to those local environments. Those features that cannot adapt succumb to those natural forces and are not passed on to surviving generations. The major natural occurrences include: environments such as too much rain or drought; severe and prolonged periods of heat or cold; abundance or scarcity of food; the strong presence or lack of natural predators; and of course, the physical attributes needed to overcome competition amongst one's own sex for mates. In broad terms it is the things that occur in a species' environment, which it has no control over, that cause evolutionary pressures and ultimately impact on that species' genome.

Somehow Darwin's sexual selection theory transcended the "tooth and claw" theory of the competing male winning a contest to be picked by a female. The theory literally evolved from the bright features of the male peacock. In fact, Darwin is famously cited as writing: "The sight of a feather in a peacock's tail, whenever I gaze on it, makes me sick" (Miller, 2000, p. 35). On the same page, Miller tells us that: "The peacock seemed to mock Darwin's theory that natural selection shapes every trait to some purpose." It made no sense: what evolutionary purpose did those fancy feathers have in the evolutionary struggle for survival? In the end, Darwin literally threw up his hands in disgust and came to the only possible conclusion: that the peahen's choice of the male's fine display of feathers was simply a matter of aesthetic choice for the female peahen. Darwin was standing at the doorway to nurture: the wondrous origins of the human mind leading away from nature's blind, autonomous reactions.

Aesthetic choice for the peahen does make evolutionary common sense. Scientists have removed some of the large dots found within the peacock's tail and guess what? The poor guy just couldn't convince a peahen that he was worthy of a mate. The theory behind the female peahen's aesthetic choice is called the handicap hypothesis, formulated in 1975 by A. Zahavi (Zahavi, 1975).

The fancier the display of male feathers, the more likely they would also attract his predatory distracters (the handicap), as well as the peahen's gaze. But if the peacock is still strutting his stuff around the farmyard, then it is also safe to assume it is the peahen's perspective that the male peacock has a sturdy set of genes because he is more intelligent or quicker at escaping predators. This then provides an outside physical clue to the peahen that the male is a very good prospect for mating because of his highly adaptive genes, which in turn could be passed on to her progeny.

To help explain further concerning Darwin's transition from tooth and claw to aesthetics, I want to again quote from Janet Browne's *Charles Darwin: The Power of Place*:

> "In animal species, he [Darwin] had suggested in the *Origin of Species*, females would mate more readily with males displaying the largest antlers, the bright-

est colours, the neatest nest, or the most beautiful song, and thereby leave descendants liable to possess the same characteristics. Over the generations such features would build up in a population. Sometimes the attributes might determine the victor in a fight for possession of the female but generally they served no life-preserving adaptive function. They merely increased the chances of mating and thus the number of offspring...He was convinced that this explained many aspects of human evolution. 'Among savages the most powerful men will have the pick of the women, and they will generally leave the most descendants....' Strictly speaking, this was not natural selection, since choice was involved. In humans, said Darwin, the choice was exercised by males. The situation was otherwise in the animal kingdom, where he believed females took the decisive role" (Browne, 2002, p. 306).

Darwin's surrender is important because it engages all of us who are trying to expand the debate between nature vs. nurture to a larger audience. It is not either nature or nurture—it is both. Think of it this way: your mind is like an astronaut taking a spacewalk tied to the mother ship, which protects you. This tow line allows you to spacewalk outside and explore this new and wondrous world around you. But with even the slightest hint of danger, the astronaut slips back to the mother ship for safety. And your mind, like the astronaut, slips back into the protective shell of Nature, responding to its natural and instinctive responses. As we evolved, our "walks outside the protective shell"—and into nurture—lasted longer and longer, expanding the storage required in our brains for all this exciting new exploratory information.

This produced a ratio of sorts of the current human mental process of protected (nature) vs. unprotected (nurture) time for our species that can be measured (or guesstimated). I surmised the ratio on today's 21st century timeline to be something like 60% in favor of nurture and 40% for nature; future ratios will be even higher pushing the nurture ratio to heights unseen. Of course when measuring this ratio we must consider the general educational level of the study group at their hierarchal level, influences of elites using media on that group, the group's social mores and values shaped by elites, and of course, the group's specific culture at that particular time and location in history.

When I first became aware of Darwin's sexual selection flaw, some of the most compelling questions popped into my mind about his bias favoring the human male: why did Mother Nature suddenly make the decision that the human male's tendency for aggression and violence was more beneficial to the survival of the species than the human female's inclinations toward nurturing and compassion? Did Mother Nature decide that in order to evolve the human species only males could make intelligent and correct choices that would best suit the evolutionary path upwards? Did Mother Nature decide that the sexual selection process that had taken millions of years to perfect suddenly needed overwhelming male bias to form a more perfect species?

Some pretty heavy questions, no doubt. But's lets back up a bit and state the fact that in Darwin's first publication of *Origin* he only hinted that humans evolved from the primates. If you know the history of events surrounding what happened to Darwin after the publication of *Origin* and the wide-spread abuse and ridicule he endured on a national level, you might understand that perhaps in order to spare himself and his family further humiliation, he decided to conform to the cultural realities of his local environment.

We will dig deeper into some of these issues as we progress through the book, but let's lay down my theory as to why and how the female led us out of the jungle.

CHAPTER 2

EXODUS FROM
THE JUNGLE

The female has an epiphany

A vast majority of scientists believe humans evolved from the primates…why would they want to leave an environment that was suitable not just for survival, but for prospering and multiplying?

A vast majority of scientists believe humans evolved from the primates. But if the populations of our early ancestors were multiplying significantly, why would they want to leave an environment that was suitable not just for survival, but for prospering and multiplying? Because most likely there was no conscious decision to leave, but either the environment changed from supportive to non-supportive, or they were forced to leave by dominating groups, or it was a combination of the two.

One theory often cited is the change in climate from a lush tropical forest to a dry, harsh savanna environment. If you accept the possibility that our primate ancestors were dominated by alpha males through aggressive behavior and the beta populations were pushed farther and away from where the alpha male was "King of the Hill," then it is very possible that those outer edges of the jungle were more likely subjected to climate change, which created a less hospitable local environment than the lush central parts of the jungle with its abundant fruits, roots, and branches.

Another possible motivating factor could be the emotional stress of the female, who was under constant sexual domination. There has been a confirmed sighting where one chimpanzee alpha male held captive a female in a tree for several days with five of the alpha male's beta male followers waiting below for their turn to share the female's sexuality. Not exactly a wonderful scenario for any female, be they primate or human.

But I posit that the truly motivating reason for the female's desire to leave the jungle would be to avoid the trauma inflicted by infanticide: in some animal kingdoms, and in a few cases of primate behavior, it is a hard fact of biology that when a male takes over from a mated female's male partner—either because that male died accidently or was killed in a heated exchange—he has been known to kill the children of the prior male's seed. I cannot begin to imagine what emotional turmoil our female ancestor felt toward the conquering male. I believe that the emotions felt were the faint beginnings of the female's thought process of not wanting to accept this choice. (Listed in the Appendix of this book is a link to a video of an alpha male chimpanzee cannibalizing part of an infant, we assume after having taken the child from the child's mother.)

So what sort of behavioral strategy could the female use to head off this terrible biological event? Would having all the males in one particular group think that they are the father of the newly born child work to the female's advantage? In some groups, primatologists have observed the female willingly granting sexual access to all the male members of their particular group, and there is solid logic behind the behavior: if the female has had sex with all the males in a specific group and a child is born, all of the males are left with the observable choice that they may be the father of the newly born child. Therefore the survival rate amongst the newly born would escalate because of the absence of infanticide.

But if the population increased at a steady rate, there would be the natural creation of more layers of conflict between alphas and betas for limited resources within a defined environment. This increased competition would escalate the natural competition for survival, which in turn would create a small minority at the top of the primate structures and increase the population of "losers" who would be continually forced to the outer edges of the local environment where resources would be less bountiful.

We know that our chimpanzee cousins have been observed using tools and passing that cultural knowledge onto their children (Goodall Institute, 2010) (and also listed in the References is a link to a video of this occurring). So it is not a stretch of the imagination for us to believe that females understood that fulfilling the males' sexual needs could be used to their advantage in an area with limited resources, and that it would diminish sexual violence or coercion as well as infanticide and its coinciding emotional trauma.

In addition, if the newly-coupled pair of beta "losers" went farther and farther away from the central core of the alpha male's territory, it was intuitively because the farther these pre-humans traveled, the alpha male's ability to influence by coercive intimidation dropped proportionately while the female's strategy of providing sexual access to the beta male or males increased. Even today we know that boundaries are very important in terms of power held by individuals, religions, and political groups (just look at

Congressional Districts for American politicians). If some of you reading this are now beginning to think that the beta males could easily be manipulated through sexual desire, you may be partially correct. We cannot escape the obvious that females selected males who enjoyed having sex at frequent intervals, for that trait was passed on through genetics. And that begs the question: was the trait a benefit for the male or the female?

This cannot be emphasized enough: "leaving the ancestral territory" is the fork in the tree of life taken by our pre-human ancestors using mental choice vs. instinctive reaction—nurture vs. nature. And it starts with a choice by the female.

How can I make such a speculative claim? Well, like a shiny apple that evolved its bright red color, making it contrast against the green background of its tree, its purpose was to attract hungry animals who would eat the fruit (and perhaps its seeds) and then discard the remaining core and seeds in a remote location, allowing the apple tree to propagate. And let's not forget the handy shape of the apple, which facilitates constant movement of a primate while searching for more food. Also notice we still have this behavior in the modern world. Humans drive into a fast-food location, purchase hand-held wrapped sandwiches, and then drive away to do other activities while eating. That, my friend, is evolutionary seed-spreading behavior 101.

If we use the chimpanzee as a model for our early primate female ancestors, we must acknowledge that they had an enlarged, brightly colored, moist, and perhaps odorous genital sac that beckoned to the alpha males. Why would any alpha male want to change his behavior to mate with a female that had a less than "perfect" estrous sac and become committed to just one female when he could just "take" the more "attractive" female of his choice?

Assuming constant conflict with beta males for sexual access to the most appealing females in their group, the alpha male would also enjoy a boost in the "male ego-boosting body chemical" testosterone to add to his inflated comportment. It is common knowledge amongst biologists that when the home town sports team wins their contest against an out of territory opponent, male spectators enjoy a 20% boost in testosterone levels. Along with this alpha male "it's good to be the king" scenario, it's also understood that there are multiple beta males wanting to replace the alpha male and enjoy the same benefits he possesses. Therefore it was most likely the female that presented to beta males an alternative to the combative route of acquiring the ultimate goal of the alpha male kingdom: sexual access. I'm suggesting that through body language the female convinced beta males that having 24/7 access to her sex was a much safer and pleasurable path than fighting with aggressive alphas.

This is not a stretch of the imagination, as Frances E. Mascia-Lees and Nancy Johnson Black have informed us that: "Studies of the sexual behavior of nonhuman primates, moreover, reveal that sexual activity outside a female's period of maximum fertility is not an exclusive characteristic of human sexuality" (Mascia-Lees, 2000). Obviously the female who allowed sexual access beyond her fecund period came out ahead with a distinct evolutionary edge. This long-term strategy resulted in the evolutionary end of an official estrous period with physical changes to the estrous sac disappearing after eons, thus allowing for true 24/7 sexual availability without visual need of the estrous sac.

Our ancestors' development of bipedalism aided their exodus from the jungle but led to other evolutionary changes. The human brain "exploded" with the multiple choices that began flooding into our memories, all of which needed to be stored in new locations in the brain. This newly enlarged brain size, along with walking upright, caused a physiological change to the primate skeleton as it co-evolved along with the brain. Bipedalism forced the female thighbones inward from the hip to the knee, which helped to balance our walking by placing the center of gravity closer to the body.

But this caused the female's birth canal to change orientation and shape. "The human female today possesses a forward-tilting vaginal canal, one designed for frontal copulation" (Fisher, 1982, pg. 95). What this vaginal canal shift did was to increase sexual pleasure for the female and allowed a view of facial expressions to the partners, aiding the pleasure of both, which in turn tempered the male's one-sided feelings of pleasure and conquest over the female. This shift was obviously preferred by the female, as it has been passed down genetically, but the shifts also increased the dangers and recovery time of giving birth for our ancestors.

> "Compared with humans, most primates have an easier time...A baby chimpanzee, for instance, is born quickly: entering, passing through, and leaving its mother's pelvis in a straight shot and emerging face up so that its mother can pull it forward and lift it toward her breast. In chimps and other primates, the oval birth canal is oriented the same way from beginning to end. In humans, it's a flattened oval one way and then it shifts orientation 90 degrees so that it's flattened the other way. To get through, the infant's head and shoulders have to align with that shifting oval. It's this changing cross-sectional shape

of the passageway that makes human birth difficult and risky…A hundred years ago, childbirth was a leading cause of death for women of childbearing age (Ackerman, 2006).

But if females led us out of the jungle, why or how did they lose their leadership? How did these intelligent beta females, after having lead the primates out of the jungle and putting our ancestors on a new path, give up or lose their leadership role and become the "second sex?" I'll give you the answer in one sentence: the size of the human brain became so enlarged throughout the millennia due directly to the female's use of choice, that her larger-brained progeny required a longer period of care, and as a result, the female needed to change in a direction that required more assistance. That assistance came primarily in the form of a taller, heavier, stronger, and more aggressive male, which the female chose.

> "One possible explanation for our slow rate of maturation is that it is an adaptation—that is, natural selection may have favored a long childhood because it had benefits that outweighed its costs. However, most scientists who have examined this issue have assumed that immaturity has no inherent advantages and that our extended period of development must therefore be a by-product of selection for some other characteristic…The most popular candidate has been intelligence. A big and complex brain takes a lot of time to develop, and in humans much of that development must occur after birth, because bipedalism limits birth-canal width, which has in turn constrained the head size of newborns…" (Remmel, American Scientist.org, May-June 2008).

Basically our female ancestor was so successful at this new thing called choice that in struggling to adjust to her new world she had to rely on others for help and had to reluctantly return to the old method of survival stored deep in her memories.

What's my theory as to the transition of the female from her leadership role back to that of subservient? That females succeeded too well through sexual choice, and like a long-distance runner running through the competition of evolution, she has merely paused for a minute or two on the Cosmic Calendar Clock in order to catch her breath before the next leg of the race. Evolution is a marathon, not a sprint. Her body and mind needed time to adjust to the equation of the longer care needed for raising children, the narrower birth canal, and the overly aggressive male she selected and helped breed. She needed more time to educate herself and to become self-aware as to what it really means to be a woman in group hierarchies. She needed to learn to become more independent, to become more physically and emotionally strong, and finally, she needed time to resume her leadership role in the world by organizing anew her sisters and engaging fellow males willing to follow them for the next wave forward.

But how long has it been, and how long before the next wave?

The Cosmic Clock mentioned earlier is the product of scientist Carl Sagan, who popularized the Clock in his successful book and TV show *Cosmos* in the 1970s (with an update from Neil DeGrass Tyson in 2014). It's the cosmic calendar, measured from The Big Bang some 13.8 billion years ago, which shrinks the time from the Big Bang into a 12 month calendar beginning on January 1st and ending at your present timeline as midnight. Our ancestors arrived on the scene at around 10:30 pm in the evening of

December 31st. It also means that humankind tamed fire and created tools at around 11:46 pm. And most importantly, at 11:59 pm and 20 seconds, they domesticated plants and animals and began to establish communities. The most amazing concept to wrap one's head around is that all the civilizations that have risen and fallen, all the wars, local and world-wide, all the kings and queens who have reigned, the religions, music, and art that have come about, and in fact, all of recorded history has occurred in the last 10 seconds on the Cosmic Calendar Clock.

I surmise that the female of our species most likely gave up her leadership role when our species switched from being hunter-gatherers to settling down and becoming domesticators of plants and animals some 9,000 to 11,000 years ago. On the Cosmic Clock that would come out to be about 11:59 pm and 30 seconds to 11:59pm and 42 seconds on December 31st. The important moment revolving around this transition would be our ancestors' ability to produce surplus food, which lead to the creation of social hierarchies separating various groups around 10,000 years ago, and which also lead to the separation of roles between genders.

* * *

In convincing the beta males to wander far from their original territory, the females escaped subjugation and forced couplings with alpha males and found greater choice in with whom and when to have sexual copulations. This great exodus, combined with searching for food farther and farther away from the original territory, advanced evolutionary pressures to stand upright on two legs in order carry provisions for long treks back to new locations.

Along with this bipedalism, the exchange of meat from scavenged animals may now have entered into the sexual access equation. Do you think if a male brought home meat protein from one of his scavenging forays his chances of finding a more receptive and grateful female increased proportionately? As a healthy, modern sexual male (although elderly), the thing I am sure of is that the male's risk in scavenging meat while other predators may still be lurking nearby could have been driven by the sexual fantasy of what might await him if he succeeded in his hunt. Sex starts in the brain and controls what happens below. Other possibilities may be that our early male ancestors witnessed such a successful exchange from others in his group or may have the stored memory from his own successful exchange.

> "With the stimulus of constantly available sex, proto-hominds had begun the most fundamental exchange the human race would ever make. Males and females were learning to divide their labors, to exchange meat and vegetables, to share their daily catch. Constant sex had begun to tie them to one another and economic dependence was tightening the knot...But those females who were able to copulate regularly— throughout their monthly menstrual cycle, throughout the entire pregnancy, and shortly after parturition—were able to maintain this economic link the best" (Fisher, 1982, Page 94).

But finding food, and therefore making this economic link, was difficult. Our early bi-pedal ancestors found themselves on the hot, dry plains of Africa where a hostile environment did not provide the fruits, nuts, roots, and small animal life in the quantities they had left behind. What most likely occurred next was that our early male ancestors, not tied to the burden of childcare, were free to roam greater distances than the female in search for food.

Perhaps he was unattached to a mate or was mated to a particular female who allowed sexual access in exchange for food, but what was new was the type of food found: meat. The source of this new protein was likely discovered by everyone noticing in the distant sky the almost daily occurrence of buzzards circling some recently deceased animal. And once one male succeeded in bringing home the prize of animal protein to his mate and her children, perhaps he shared with his fellow males what little food and knowledge he had of how he located the new food source with grunts, voice intonations, and "pointing" to the sky to indicate the circling buzzards.

And let's be honest here: most likely our male ancestor was a desperate hungry male. In the early years after migrating out of Africa, he was most likely not a hunter but a scavenger, picking the remaining meat from the outside of the bones and the marrow within from a prey felled by some other animal predator or from the occasional gift of an animal felled by accidental providence. The consumption of meat protein has been cited time and again as one of the major contributing factors for the human brain exploding in size. Eventually, by imitating the success at getting food alone, successful males were joined by other male imitators as they ventured out individually but then recognized each other from their same grouping as they arrived at the distant buzzard circling locations.

This early coincidence of males arriving at the same time separately may have had a logical outcome: if all the males knew from sight that the other males were from the same tribe or territory, would they throughout time have developed male-bonded groups venturing out together to the circling buzzards with a common objective? If there was a frequent, complex movement between the males, it was most likely repetitive grunts, raised or lowered tonal verbal expressions, and body gestures developed into a more complex form of early language and male bonding. Don't worry ladies, I haven't forgotten your skills at language, multi-tasking, and gathering, but let me dwell on the possible formation of patriarchy in its early stages.

As populations grew from these successful groupings of early "families" with their males venturing out together farther and farther, they probably met males from other tribes or groupings searching for the same food source. And if so, there were most likely physical conflicts over the remaining scraps of meat left on the bones of fallen prey. Yes, most likely there were conflicts—and, more than likely some of these conflicts ended in injuries and death. So did the male winners after the physical conquest "pound their chests in victory" and the losers slither away in dejection or perhaps show deference in the manner of our primate ancestors? Interesting questions and possibilities.

With the of expansion of verbal organization, early man then plausibly traveled in groups to give each other courage and safety, which over the millennia helped to establish even further this complex yet important male collaborative cohesion, which will bear heavily in the future as a male-grouped bias against women in the form of culturally constructed instructions and knowledge.

Now let's jump ahead a bit and look backward at the close human behavior of sharing personal knowledge today. This is a one-on-one culture, which is handed down from generation to generation and from one person to another, and the receiving person trusts the giver implicitly. It is a powerful method of exchange, a method recognized by advertisers in our modern societies. It's also the basis of gossip, but that is entirely another venue in the survival game that we will briefly discuss elsewhere.

Both sexes use this method of gaining valuable information about the opposite sex in the mating game. Both males and females will develop their own mutual knowledge and understanding of the opposite sex's physical, emotional, and sexual appetite and pass it along to each other verbally in the same manner with the same powerful results. Both genders pass this information on to each other as accurate, and of course it is mostly anecdotal, but that matters not in the phenomenon of evolution, where survival is the name of the game. Trust of the information received is contingent on the familiarity of the person giving the information to the recipient but also on the desire on the part of the recipient wanting to know the information received.

Another important factor in the future of male behavior would be this constant wandering of males—the movement beyond the base camp—for the continual gathering, scavenging, and eventually group hunting for food. This dangerous roaming away from base camp into the unknown undoubtedly led to a thinning of the male population based on those who survived vs. those who failed to return from the hunt whole and uninjured. It would also lead ultimately across history, influencing males with an itchy-foot syndrome: the desire to explore beyond their territories in search of trade, treasures, and military conquests. And unfortunately, elite rulers in our ancestors' future have used this male wanderlust in combination with the evolutionary success of over-population, along the way defining bachelor males as expendable tools for territorial conquests.

What do you think our early females did when their mates did not return from the hunt or returned injured so that they could no longer hunt and help provide meat to the children? Was the female faced with a stark reality of not being able to gather enough food for herself and her progeny, thereafter deciding to choose a gift from another male in exchange for sexual access while her injured mate lay nearby to watch the sexual exchange? If so, how do you think that made the non-providing male feel? Did the new male empathize with his injured comrade? Or was his need to satisfy his sexual urge so strong that it did not matter how his injured friend felt emotionally? If our early primate sisters faced these difficult situations, can you answer these questions by establishing what personality traits survived the journey across the African plains and were passed onto her progeny?

What choices tore at her emotional fabric? Were the emotions of separating from her established mate tearing at her if and when she crossed over and accepted food from a potential new mate? The choices available were obviously not simple, but ultimately it always comes down to preservation of the children. Evolution for the human animal is not about survival of the fittest. It's about who creates and leaves behind the most children by deciding how those children will survive, by making choices based on innate sensory perceptions of the local environment, and through nurture-based knowledge (let's call it the early stages of culture) founded on group cultural knowledge.

Enter the early human emotions of hate, anger, fear, guilt, depression, and disgust—all based on biological functions found deep in the inner core of our brains—each expanding outward and finding new ways to be expressed so that others near us could empathize or sympathize with those emotions being expressed, an essential element for survival in our newly expanded world. And don't forget, these new emotions needed to find room in a mushrooming brain to preserve the circumstances surrounding the moment, as well as new architecture to express outward those new behaviors chemically stimulated by emotions as part of this new emotionally based world. Since all of this new information was and is important for survival, it was passed on to succeeding generations.

And as populations grew exponentially because of acquired knowledge and skills, young males and females growing up in extended family groups, tribes, and villages began to see the advantages successful male hunters had. These males were larger and physically dominant over other males and females. It's possible these young hominids saw hierarchies form among successful male scavengers (now male-bonded group hunters), winning greater success with fecund females along the way, and likely they also witnessed females succeeding in having more children by becoming dependent on the most successful hunters. Surely they saw females accepting gifts of food and witnessed events where the larger, more aggressive males were able to protect the females and their children.

Before language, views and knowledge were passed on by visual experiences. If there was a "language" to make these views known, perhaps tales of bravery were re-enacted and they turned into exaggerated myths, told around the new invention of cooking meat protein around a campfire. Perhaps myths reflected the physical ideal of the male provider and protector of the tribe as the tallest, strongest, fastest, and bravest male hunter and were tied directly to the survival of the tribe. Most likely these exaggerated, unreliable tales, passed on from generation to generation, begin to give the male an exaggerated sense of his own self-importance.

> "The larger male protohominids probably roamed with more confidence, explored farther afield, and found more animals in their wanderings. This made

them better food providers, which attracted women. Perhaps they were better fighters and better protectors too. In any case, if females preferred large, strong males, then their genes would proliferate, producing the larger human male we see everywhere today" (Fisher, 1982, pg. 96).

It is my belief that during this point in our human history, after succeeding generations of our male ancestors had become larger, stronger, faster, and braver, besting other males at scavenging fallen prey, and perhaps even believing their mythology, the males also had an epiphany. The male saw an opportunity to dominate due to his physical size, and so he also made a choice. He dipped into the past behaviors of physical violence or the threat of violence in order to gather greater resources and to attract a female mating partner. This new male discovered that his strength also gave him easy domination over the now physically weaker female who, remember, had chosen that path genetically because it aided in attracting through 24/7 sexual access the strongest, bravest male with the most resources. And how did the male come to this conclusion?

Undoubtedly new generations of young assuredly observed that these new "supermen" were attracting the more accessible, or the more sexually appealing, fecund females. Was the female returning to the manner of her primate ancestors in seeking the behavior of the alpha male back into her life? Well, somewhat, at least—as our female ancestor's may have been using the innate knowledge of the power that her sexuality has over males. But it's not quite the same story because there is no alpha "king of the hill" to compete with, as highly competitive male primate behavior had been greatly reduced or eliminated in the egalitarian cooperative journey out of the jungle and the essential need for cooperation amongst migrating groups.

That is the core building block of culture. What may seem as trivial events, people, religions, or social structures found at every location on the planet at any particular moment in time turn out to have this core foundational behavior of survival influencing choices with sexual coupling as its base. What primary change has occurred is that mental choices have influenced our pre-human ancestors and widened the separation between nature to nurture, allowing individuals to make decisions for themselves. Yet those decisions are still heavily influenced by biology. It is the genetic code in our selfish genes wanting to perpetuate themselves and allow this biological vessel the opportunity to make directional changes based on successful outcomes from the passage of genetic codes.

As for the female's role in leading our ancestors out of the jungle, there is some scant evidence to help confirm this theory, so at the very least it is speculative. Yet there is still evidence: "The idea that human society was originally a matriarchy with female deities and female leaders was taken up by a few archeologists studying prehistoric cultures in Europe, most prominently, by Marija Gimbutus. Gimbutus argues that during the Paleolithic and Neolithic period, people living in Europe and the Mediterranean area were egalitarian, peaceful and woman-centered, honoring the earth as a mother goddess. This "Old Europe" was gradually overtaken through conquest and migration after 4000 BCE by Indo-European speaking people who originated in the steppes of Russia. These new people were militaristic, semi-nomadic, and patriarchal, and they worshipped a single male god and often followed a single male military leader" (Wiesner-Hanks, 2011, pg. 17).

Volumes could be filled discussing the presence of the Divine Mother in all cultures from every corner of the world. In these earlier mythologies, the Mother was seen as that which gives birth to all creatures and that the earth and the elements were not void of spirit but are in fact the living Goddess-Creator herself.

> "The Goddess gradually retreated into the depths of forests or onto mountaintops, where she remains to this day in beliefs and fairy stories. Human alienation from the vital roots of earthly life ensued, the results of which are clear in our contemporary society. But the cycles never stop turning, and now we find the Goddess reemerging from the forests and mountains, bringing us hope for the future, and returning us to our most ancient human roots" (Tate, 2006).

The possibility that an entire cultural group could be wiped out physically by another dominant group can be posited with historical accuracy. The most prominent evolutionary example was the mysterious disappearance of the Neanderthals. Once thought to have been completely wiped out, recent DNA evidence proves they had instead been absorbed by our human ancestors. There of course was the Spanish conquest of the Aztecs and the American Indians being shoved into territorial reservations as white settlers moved west. One of the most blatant acts would be the passage of a law called the Indian Removal Act of 1830, making the destruction of an entire ethnic group legal. Historically, in the American South up to the 1960s, African-Americans were routinely lynched and sometimes lit on fire to the cheers of gathered whites in carnival-like settings. And let us never forget the "final solution" of the Nazis attempting to eradicate the Jews.

Were these atrocious behaviors the result of our passage from the jungle to civilization merely all male behaviors, or was the female a subservient, silent collaborator? If it was possible that an entire race or subspecies of "peaceful" feminine-led hamlets or pockets of wide-spread villages could be almost wiped off the face of the Earth, would the few surviving females rethink their plans to exchange the sexual access that would lead to more aggressive defensive behavior required from males? I sense that must have been the reality at hand.

The fact that little remains in the archeological records as solid evidence also walks hand in hand with hierarchical social misogynistic tendencies of males—with support from elite females—to suppress any evidence of feminine superiority for their own group and familial advantages. This suppressive tendency to wipe out the female goddess does make some social hierarchical sense when viewed in context with the Social Dominance Theory, which we will explore in Chapter Eight.

There appears to be a general agreement that our ancestors left the jungle in cooperative egalitarian groups, and that at some point in their journey they came upon a fertile area that was lush in vegetation and water. But I speculate that the female just got tired of being on the move, trying to follow the beast that they had previously relied upon on their way out of Africa. It's also possible that settling down and growing food and domesticating animals instead of chasing migrating animals constantly made it easier for the males to protect the females and their children—and to make more babies.

This theory that the female led the male out of the jungle by use of sexual temptation is not too speculative. Females today make every effort to keep other women away from their prized boyfriends in school or their easily tempted, philandering husband. Why? Because the present female of today wants to reap the rewards that their female ancestral parents worked so hard to achieve in order to protect their investment in keeping a committed male around to help propel their progeny into the next generation.

And, as we will discover, the invention of the plow causes a greater shift in female dependence on the stronger male. And with this greater dependence comes a cost to the female.

CHAPTER 3

THE PLOW

The male has an epiphany

"... There is probably no single tool in human history that wreaked such havoc between women and men or stimulated so many changes in human patterns of sex and love as the plow..."
(Fisher, 1992, pp. 278 & 279)

Along with this conceivable shift in the female's competitive approach against her fellow sisters, was there a sudden increased need to seek out protective males as well? Or was it the other way around? By this time in our evolution, it was the male who brought back scraps of meat, and who was then fortunate enough to be granted sexual access. So logic dictates that in the male world, success was about competing to be chosen as the male worthy to assist the female in the survival of her progeny. In a world that had been flipped upside down from egalitarian to one that was resource-limited, did the male find that he was thrust into a more prominent position in the mating dance? Or was it a combination of both genders focusing on survival at the specific moment in time, which set off a series of behavioral shifts? If once plentiful resources suddenly disappeared, would the original non-violent and cooperative interactive behavior change to one that was more competitive while still maintaining the ultimate goal of survival of the progeny at all cost?

Whatever the direction brought on by the local environment, obviously our ancestors survived. But a confluence of events was emerging from these successful farm settlements that would shape human behavioral patterns for generations to come. As populations continued to grow, small groups were maintaining more resources than others, simply because of their own hard work—and of course just plain dumb luck of planting in a particular location with better weather and soil conditions. Anything left over from consumption by the family unit would be considered a surplus and was stored. These stored up resources consisted of such items as dried meats, fruits, roots, hunting apparatuses, protective clothing, adequate housing, domesticated animals or plants, etc. Now, understanding that a resource is anything that will help pass a species' genes into the next generation, what logically happened next was that some individuals and groupings of individuals had more resources than others. These resource accumulations are all associated with survival of the species and have taken on major importance in our minds as a "comparative balance sheet."

This balance sheet of comparative resources appears to have evolved from our early animal origins and were once based on physical strength separating individuals ranked into hierarchical positions. But once our ancestors became successful at the early stages of animal and plant domestication, those with more "stuff" were viewed differently in comparison with those who had less stuff, which is fused with the mental association of these resources as a biological basis of survival. This is important because this hierarchical positioning is no longer based on strength or coercion (nature) but on mental comparison (nurture). What is important to note: we could be at the mental emergence of group identity associated with pride, envy, and greed.

Today we would call these resources by many names: salaries, dual-incomes, inheritances, as well as monetary values placed on one's dwelling, savings, potential earnings, pension funds, vacation homes, the ability to pay for education, investing in children's college education (in advanced countries); and don't forget cars, clothing, jewelry (for men and women), and the type of phone/pad/computer they carry. And those are the just the physical items with which we can surround ourselves. Then we throw in our own aesthetic attributes dictated by socially construed influences, which have a profound effect on our social western world—particularly young, sexually active adults. If you are male of 6'5" and weighing 250 pounds of muscle in the proper proportions instead of 5'6" and 250 pounds of fat, you are a more sexually appealing prospect to a physically attractive fecund female with the hip to waist ratio best suited for reproduction most appealing to men. Put all these culturally dictated attractive reproductive signals into a human package that has an outward positive mental attitude towards life and interaction with others, and you have the basis of a "winning reproductive package." Everything is based on the dominant culture at a particular location and time in history.

These are all physical and aesthetic resources that help to propel individuals and family units genetically into the future. Inside our bodies are codes of instructions— tucked away in neat little packages—carrying the outcomes of all the successes and failures of previous generations guiding us on our evolutionary voyage. In our pre-human past, all these instructions were automatically decided for us biologically; once again, we call this *nature*. Now, because our brains have expanded throughout the eons, we have added choices to the genetic mix to guide us in life, to allow free choice, or what we call *nurture.*

You may recall I mentioned that important new phrase: family unit. With family units, it is not just one being trying to survive; it is a combination of small family units and unattached males and females trying to survive in an established local environment. Of course I use the phrase *family units* very loosely, as there were no institutions with religions dictating morality or values yet; I suppose the more accurate word would be *conjugal units.* We are now at the point in our early history where our ancestors were not just forming small mated pairs and small family units, but belonging to a tribe with established territory and established houses, be they huts made of animal skins, mud, or sticks and stones. What is important is that they no longer lived in caves because they had discovered fire, and smoke made living in caves unsuitable. Moving from cave dwellings into the open exposed our ancestors to the elements, so new dwellings were devised. It is also the time in our evolution for making rudimentary tools, and most importantly, tools for speeding and spreading the domestication of plants, such as the invention of the plow.

Most historians and anthropologists believe civilization began in an area called Mesopotamia. In today's boundaries, the area would combine Assyria, Babylonia, Jordan, Iraq, Lebanon, and Palestine. In Greek, Mesopotamia means "between two rivers"; those rivers are the Tigris and Euphrates. In the northern part of this area there are rivers and streams fed from mountains, and some speculate that the area was filled with wildlife and edible vegetation. In his Pulitzer Prize-winning book, *Guns, Germs, and Steel: The Fates of Human Societies* (Diamond, 1997), Jared Diamond theorizes that Western Civilization was mostly luck. People thrived on a perfect combination of mild weather, fertile soil, and the domestication of plants and animals, then they migrated west and northerly along the 34th Parallel, where these soil and weather conditions were optimal. Into this mix of optimal agricultural conditions, following the biological female birth canal anatomy changes requiring longer-term child care by the female and her developing preference for stronger and in-bred aggressiveness of the male, we arrive at the moment in history marking the male's mastery of the plow and the early stages of patriarchy.

The most important point to be made about the plow is that, along with the domestication of animals, it created surplus food beyond mere daily existence. This surplus transformed the equality quotient of the hunter-gatherers' means of survival and created family units that no longer required—and in fact perhaps rejected—the meme that their hard work had to be shared with others. This selfish choice evolved because the family units were no longer on the move from one local environment to another where hunter-gatherer groups were in the same boat for survival and therefore depended on each other. Think back to our ancestors being pushed to the outer edges of the alpha males' territory, which is basically a specific area that had year-round resources. Early agricultural surpluses as a direct result from the invention of the plow can be considered the first accumulated wealth within a specific local environment, which in being separate from those more lush, bountiful areas on the planet were then subjected to fluctuations of good and bad years of food production. It is a pivotal moment in human history within which our ancestors created their own resources vs. relying solely on nature while remaining in one location.

There does not seem to be a definitive moment in recorded history as to when the plow was invented or what form it took; different studies have the invention date in a wide span of history between 6,000 BCE to 3,000 BCE. That seems like a long time to us shortly life-spanned humans, but on the Big Bang Cosmic Calendar Clock it represents a mere six seconds. In all likelihood it developed through trial and error attempts at planting seeds by crude hand tools, and then intuitively evolved as a way to do things better and faster. On the positive side, the plow was important because it increased food supplies, but on the negative side, the device required hard, labor-intensive work in many climates that sometimes were not ideal.

In fact, some historians now have called this time period of the plow's development as the "plow agriculture" era, which resulted in further division of labor between males and females but with a new, major social development: there were some families who owned plows and some who did not. The importance of the plow cannot be overstated, as it helped to produce more food, which meant that the surpluses could be traded for other items. It also created the need for storage and the need to record any trade transactions, thus creating further need of language and mathematics to tally up those resources and the trade differences. The storage of surplus grain also required new structures to be built, along with the knowledge needed to build those structures, and a new added role for males: the need to defend those structures against raids by rival tribes or roaming bands. This led to the rise in need for the creation of further weaponry. These early agrarian tribal villages that survived and prospered reasonably grew into cities and then city states.

"…There is probably no single tool in human history that wreaked such havoc between women and men or stimulated so many changes in human patterns of sex and love as the plow...In cultures where people garden with a hoe, women do the bulk of the cultivation; in many of these societies women are relatively powerful as well. But with the introduction of the plow—which required much more strength— much of the essential farm labor became men's work. Moreover, women lost their ancient honored roles as independent gatherers, providers of the evening meal. And soon after the plow became crucial to production, a sexual double standard emerged among farming folk. Women were judged inferior to men" (Fisher, 1992, pp. 278 & 279).

Of course my conflict with the above statement would be that the strength needed to plow was a trait selected earlier by our female ancestors in the male. She chose these strong traits because of the greatly increased time needed for childcare, directly brought on by bipedalism and the increased size of the human child's brain. However, Ms. Fisher's observation that "soon after the plow became crucial to production, a sexual double standard emerged" is spot on; this event led to the creation of comparative wealth, but it also created something disturbingly negative: the view of the female as "weak" and "useless."

"The first written evidence of women's subjugation in farming communities comes from law codes of ancient Mesopotamia dating from about 1100 B. C. where women were described as chattels, possessions...Unlike women in nomadic foraging societies who left camp regularly to work and brought home precious goods and valuable information, who traveled freely to visit friends and relatives and ran their own love lives, a farming woman took her place in the garden or the house—her duty to raise children and serve a man…With plow agriculture came general female subordination, setting in motion the entire panorama of Western sexual and social life" (Fisher, 1992, pp. 279—281).

I believe we are near the moment in history when our male ancestors noticed and understood that there was an advantage to being larger, stronger, and more aggressive than other males; it gave them a competitive advantage in gathering resources learned in the scavenging and hunting stages of our human history. Like our female pre-humanoids who saw an opportunity to change their world through choice vs. nature by guiding us out of the jungle with unlimited sexual access, the male unquestionably also saw an opportunity in that his new physical traits could be used in the gathering of resources, which in turn would garner him more sexual access. I posit that our early male ancestors also intuitively had an epiphany: that the use of aggressive behavior was advantageous in gathering resources to attract females who intuitively knew that these resources would give her progeny a boost over other children. In addition, now that our ancestors were no longer migratory, this aggressiveness was also focused on one specific location that had to be defended against other aggressive males. It was at this place in time that early constructed culture within specific territories emerged, with the overriding meme that the female was useless in defending those specific territories so vital to the survival of the family unit. Now compound this with the female's general physical inability to operate the plow—probably invented by males—and we have the combined elements of a perfect storm of constructed culture gathering against the female as an essential element of the family unit.

But keep uppermost in your mind that sexual access is the common thread throughout history, and remind yourself that at this moment in history and for many years to come, that there is no knowledge of the internal forces of selfish genes (Dawkins, 1976). Early patterns of patriarchy emerged because of the plow and the food surpluses that were produced, and that, I believe, primarily led to female infanticide rates found today in rural parts of Earth where customs evolved from the agrarian past (Hudson, 2004). [Also see Appendix for film links.] What's missing, of course, is that the majority of women accepted this role reluctantly only because when options are reduced to zero, survival through choice boils down to the least horrible choice. I know, I know, ladies—I'm just another masculine jerk reveling in my superior position. This is the path evolutionary pressures placed before our ancestral sisters—it's not written in stone that this is the path females will take in the future. These early enemies of feminine equality, who selected aggressive males for survival, are the females of our modern world who accept these restrictions—either consciously, or unconsciously—and know innately that they benefit her children over some other female's family units in lower social hierarchies. The following sentence from a female anthropologist is directed at female readers, and perhaps to all who understand social hierarchies and gene transference, explaining the female's evolutionary determination: "When push comes to shove, it will be my progeny that get passed to the next generation—not yours" (Campbell, 2002).

The ironic outcome of this evolutionary psychology for women, of course, was that it was their children that reaped the advantages of their mother's overriding progeny advancement behavioral mechanisms. Those benefits were passed on to the free-thinking radical feminist sisters of the first wave of feminism in England and continued through the American 1970s movement, leading me to surmise that, yes, the female is in control of our evolutionary path, but at different hierarchical levels, and at various moments in time. I come to this hierarchical emphasis because of the actions of some American women to defeat the ERA (Equal Rights Amendment) in the late 1960s, emphasizing that equal rights for women would be a demotion. The only conclusion to reach would be that high-ranking females would consider this to be a demotion through the rising up of "inferior others" to their level, which would then greatly reduce or remove any advantages they now possess.

We will explore these extremely important hierarchical levels and the human behaviors surrounding them as we continue on our journey through time, but we still need to focus more on this male epiphany and his ascendancy using force and mental degradation directed at the female. I will argue that the essential function of reproduction— the female's menstrual cycle—along with the invention of the plow contributed to the growing cultural construct among men that women were inferior.

> "Plow agriculture increased the amount of food available, but also increased the amount of goods needed to produce and store that food, including animals, storage bins and containers, and wood or metal equipment. This economic gap between families that owned lands and plows and those that did not—widened, and social differentiation increased" (Wiesner-Hanks, 2011, pg. 61).

The above paragraph nicely sums up the general path our ancestors were on: that hard work, in combination with the dumb luck of finding the best agricultural locations, helped some families succeed more than others. This difference, according to Wiesner-Hanks, produced an "economic gap between families." I think that it is important to point out before your mind drifts too far into the world of modern economics that we rephrase the paragraph to redirect your thoughts back to the evolutionary environment surrounding our early ancestors: the accumulation of resources and the behavioral mechanisms required to maintain those resources were seen as a genetic advantage while living in close-approximation of other family units in specific local environments on the planet at a specific time in history.

Also recall Helen Fisher's statement about the plow that it "wreaked such havoc between women and men." This havoc Ms. Fisher wrote about is the split from egalitarianism to a more competitive nature. The important question to ask at this moment in history is this: did the male begin to believe his efforts were more important than the females' ability to reproduce? I argue that the male's delusion of greatness that came from using the plow and aggressively defending his village led to cultural beliefs fabricated by them throughout history that the female was weak, inferior, and useless, and that these beliefs continue to blight the gender landscape up to this very day.

This derogation of females as weak and inferior has a long history, and I believe that a good portion of this attitude emerged from fabricated mythologies associated with the potency of the female's menstrual blood. There are many myths created by early male hunters that still exist today in isolated areas, such as: if a woman touches a man's spear during her cycle that the prey would catch a whiff of her blood and spoil the hunt or perhaps put the male in danger (Knight, 1991). Myths have an enduring life of their own and acted as powerful, everyday guides that helped humankind sustain itself. Unfortunately, they also can anchor our evolutionary voyage, keeping it from advancing for many years.

CHAPTER 4

ONLY WOMEN
MENSTRUATE

The male expands dominance and control

"If men could menstruate…Military men, right-wing politicians, and religious fundamentalists would cite menstruation ("men-struation") as proof that only men could serve in the Army ("you have to give blood to take blood"), occupy political office ("can women be aggressive without that steadfast cycle governed by the planet Mars?"), be priest and ministers ("how could a woman give her blood for our sins?") or rabbis ("without the monthly loss of impurities, women remain unclean")."
(Steinem, 1978)

There is one undeniable truth about female biology: women menstruate so that our future species will continue. This is important to discuss because the impact of this monthly event on every female who is still able to reproduce affects the female emotionally and socially and, I strongly suggest, served as a contributor to the emerging anti-female bias that surfaced shortly after the invention of the plow.

With hormonal changes each month, a women's body builds up a thick layer of tissue lining her uterus in preparation to receive and provide nourishment for an egg that leaves one of her ovaries and enters the uterus. If the egg is fertilized by the male's sperm, her body goes through additional hormonal changes that support the growth of that two-celled zygote into a human being. If the egg is not fertilized, the female's hormones that built up the lining in preparation for nurturing drop, causing the unneeded tissue including the unfertilized egg to exit her body in the form of menstrual blood through her vagina; her body then begins its monthly cycle once more.

Much of what roiled around a woman's menstrual cycle was a mixture of religion and mythology, and all of the beliefs that spread throughout the culture were without any basis in fact. The outcome of all these myths and cultural constructs was a disrespect of women by men, and the physical distancing and exclusion of her from activities in religious services, society, and leadership in any endeavor.

Most of the myths had no basis in science but were powerful because of their wide-spread usage, and unfortunately, the general acceptance by the female populations themselves. But these attitudes can be attributed to the lack of proper education systems in place amongst male-dominated societies. The two most prevalent myths were that females could not approach their deity in a religious setting because they were unclean and, as mentioned previously, could not approach a hunter before he left on a hunt for fear of causing "bad luck" to the hunter if she touched his spear. The common thread in hunting is that myths created around menstruation are bad for women and uphold the male myth of superiority—especially that of a male hunter (Delany, 1976 & Houppert, 1999). Do you recall the basic reason for Darwin's Flaw surrounding hunting?

It should be noted that some scientists believe the segregation into "menstrual huts" was created by women to afford them a break from the tedious, mundane duties of early farm life and to give themselves a chance to socialize (Buckley, 1988). However, I equate these menstrual huts to the "separate but equal" concept created in the American South, where the dominant white population was able to maintain segregated schools, swimming pools, and drinking fountains in order to uphold their cultural beliefs and continue turning them into advantages for their children. Hence, it is my position that the menstrual huts were socially constructed separation areas primary originated by males and were used to uphold their "superior" advantages.

> "Thus the institutions of English patriarchy, inherited from Hebrew and early Christian societies, rested upon twin pillars: the subordination required of women as a punishment for Eve's sin, which was fundamental to biblical teaching, and an understanding of men's and women's bodies, evident among early modern medical writers...in terms of relative strength and weakness. Patriarchy was thus founded upon God's direction and woman's natural physical inferiority. The crucial link was that between the weakness of mind and conscience which caused the first emblematic act of disobedience and the weakness of body which, connected to their reproductive functions, deprived women of a constant public role" (Fletcher, 1995, p. xvii).

Since we believe that the separation of the genders may have quickened with the heavy plow and division of labor, I also believe that the division of males and females further accelerated when religious ceremonies were just being established. Because early males intuitively were blind to the general innate force behind reproduction, knowledge escaped early male intelligence and somehow got twisted through history into a negative bias against the female.

I strongly believe that the early bias against women was accelerated by males attempting to assert that their god was male and that the female's menstrual cycle made her unworthy to come before that god, claiming that this essential monthly event made her "unclean."

We must also acknowledge that water was in short supply. This reduced for women the chance to become clean through frequent bathing. In addition, there were no modern products to absorb and collect menstrual blood, nor were there any chemical perfumes to disguise the odor (at least not for the common poor female of the time).

We must remember that the emotion of disgust is a biological warning sign developed through evolution to warn the body of possible dangerous substances—primarily through smells. But what the mind chooses to do with that information is no longer considered a contributing factor in *nature*, but now primarily *nurture*—that is, culturally constructed. How it is acted upon is then passed culturally throughout the specific location and time on the planet. If the action continues from generation to generation, it then becomes a meme—which, in this case, unfortunately becomes this anti-feminine bias heavily dominated by male religious dictates.

There is no doubt that these male-dominated religious cultures caused early gatherings to be segregated by gender. I posit it was because of the biological emotion of "disgust" being attached to the female's menstrual cycle, and that the male-dominated religious leaders used that as evidence that women should not be allowed to be near the holy place. If females were forced to be segregated at the same time that both males and females came together to worship their god, then this would have attributed to a sense of shame in females in the aggregate. Shame is another emotion arising from the brain's consciousness. It is a painful sense one feels when they've made a mistake in their actions and need to correct the behavior in order to be accepted within their social group; shame is attached to the sense of inferiority in one's self, and that is a socially constructed meme leading to a consensus within the culture. The culture in this case would be that males and females believed menstruation was shameful and made her unworthy to be before her god. This shame seems to have spread to all social events in the female's life, causing her to be hidden away during her menses.

The answer, of course, is education and the reversal of the negative culture revolving around this vitally important biological monthly event. I cringe in disbelief at the lack of understanding of the cycle and the distancing behaviors still used by our species.

You may be thinking that the date on our history calendar is somewhere around 3,000—2,000 BCE and that our early ancestors knew very little about the female's reproductive cycle, and therefore it would be easy to understand how myths and falsities got started. But what about now?

To find out, let's jump forward 2,000 years of human intelligence and progress to the year 2012, and to India, as young girls learn about their menstrual cycles.

> "The girls…were…told: when you menstruate, don't cook food because you will pollute it. Don't touch idols because you will defile them. Don't handle pickles because they will go rotten with your touch. My mother says it is so. Her mother believed it, and her mother before her. It must be true…Girls suffer if they aren't empowered to manage their menstrual cycle without pain and shame each month… Their health, schooling and dignity are in the balance…And the world suffers too: educated women

are healthier, have smaller families, often earn more and have a positive impact on development...it can take years, even generations, to change a taboo...But anecdotally, outreach workers note that the only girls who don't believe the superstitions about menstruation are those with educated mothers. So the best way to change the minds of future women is to keep girls in school today" (George, 2012).

So one of the solutions is merely to educate all young females around the world so that they can pass on facts, not myths, to their future progeny. Easy to pontificate the solution, but difficult to implement. Why? Because it involves elites in their specific societies who can put educational programs like the one discussed into place yet who have apprehension that resources will be taken from them, which in their view would lower their evolutionary advantage over those lower (read: poorer) levels in the hierarchical layers. This resistance by the controlling elites against perceived loss is based on both nature and nurture.

The above excerpt from an article from *The New York Times* is a bit scary because of the timeline, but it does give us an easy peek into our past, and we can see that the evolution of our species takes a weaving side by side yet still forward and upward motion in the tree of life. As you'll recall, I mentioned the passage of important, especially personal information from one trusted individual to another (in this case, from mother to daughter), and that it has a powerful effect on the recipient of that information. Using this 2012 article, we can reasonably assume that in our early history incorrect information regarding females' bodily functions was passed along, and that these "facts" could have been used in the creation of a negative bias against the female.

So the overall premise follows that these early cultural myths, where the female was weak and broken, were beginning to take form in early hunter-gatherer tribes and could have been manipulated by the male who may have had an epiphany about his relationship with his mate.

As surplus foods became more available and the successful settled populations grew beyond the basic need for survival, this obviously grew the various family units within a specific and semi-permanent location. What we had then are basically small agricultural families working on small farms with the male and female working side by side, usually located on the outskirts of an emerging city that sprang up along a trade route by land, sea, or river. What it also created was a shortage of fecund women because the major cause of death in those early years was giving birth to larger-brained children. It was, once again, because of all the successful traits that were transferred onto the succeeding generations while the physical body of the female had to adapt or perish.

Plausibly this produced a greater population of bachelor males. Failing to find any suitable mates close by, they would travel to the countryside and migrate to other villages.

Here's the problem for the female: one of the traits passed on to males was the competitive nature associated with the seeking of sex. The act of sexual intercourse is so pleasurable that the males' need to release their semen built up over a short period of time, for some men, is almost a necessity. Mother Nature is so adamant that male's release their sperm that even if the male does not release his sperm consciously, the sperm is released during sleep through a nocturnal emission commonly known as a "wet dream." As males progress in age, the messy nocturnal emissions are avoided through self-stimulation—masturbation.

But masturbation is only a temporary fix. There still is the mental substitute for the real biological imperative of reproducing the "natural" way—sex with a female. This, of course, requires the male to find a legally, morally, religiously, biologically and socially suitable place for the sperm to be deposited. And unfortunately, some males are unwilling to travel down the straight and narrow path that most females' desire and social mores dictate, and this results in culturally unacceptable behavior such as rape. Rape is not about sex; it is about domination and control. Recall the first chapter, where it is the alpha male forcing sex upon the female, and the female is deprived of choice.

With this excess population of unmated males (either through death via childbirth or the infanticide of females infants), any surviving female (but usually fecund) would become a target of possible insemination. So the male who currently has access to and is mated with a fecund female must now defend his mate against possible forced sex by other males, and hence, "mate guarding" arises.

Adding to this dilemma is the knowledge of culturally astute females that it is easy to seduce some males to do their bidding through, in some cases, the temptation of the offer of sex. This now poses a concern to the mated male as to whether his children are his and not the product of some other male's genes. If this occurs, the theory goes, the male would waste precious time and energy raising another male's genetic code. In the common cultural lingo this is being cuckolded. So now we can begin to see the culturally constructed behavior of the male with the combined evolutionary pressures to assure positive proof of paternity and the need to guard against false genetic mistakes. In simple language, it is a nature and nurture seeking common agreement. Now, if you're a male who is larger, stronger, braver, and more aggressive, what behaviors would you expect the mated male to do to assure that he will not be cuckolded?

Mate-guarding basically means that the male physically guards his mate against any outside sexual access. It's easy to see how this mate-guarding could easily slide into new word usage and a twisted belief that the female is "property" owned by males.

Today there are dozens of web sites that detail all the abusive behavior by spouses (mostly male against female), describing verbal intimidation, humiliation, and threats. Included are the behaviors designed to rein in the spouse such as monitoring and controlling every penny that the spouse spends, actual withholding of money or credit cards, and the ultimate abuse—isolation, the actual cutting off of their partners from the outside world, from friends and family.

So the theory emerges that these modern spousal abuse behaviors evolved from the havoc created by plow agriculture, leading the male to decide that the female was useless as a farm laborer due to the much longer time required for child raising, the religious myths originating around the female's "original sin," and the myths arising from the female's menstrual cycle. However, the biggest reason for males isolating females was his knowledge of his own high sex drive and that it was present in other males. This justified his mate-guarding behavior which in turn protected his home, property, and warded off cuckoldry. In terms of economics, this turned the highly valued sexual access of the female into a scarce commodity and drove up the price of that access. It is at this point in our evolution where males could easily conceive the isolated female as property alongside his house, barn, plow, and farm animals.

The ironic outcome of all this, and the hardest to fathom by feminists, is that these hoarding and aggressive behaviors by males have been selected by prior females as the trait most desired to advance their progeny when resources became scarce. And there is a difficult task ahead for our society if we wish for positive changes in those societies. These inborn behaviors are not like the remote-control on your TV—they take generations to form, be perpetuated, or die out. And those in certain high social hierarchies will resist any changes because these behaviors are a social and biological advantage to their progeny. The advantages of these behaviors become clearer soon as we explore social hierarchies.

We can condense the reasons for the rise of patriarchy up to now into a four-legged stool:

- The female-selected tall, strong, aggressive and courageous males due to necessary extended child-care requirements; selection is mentally selected and is thought to be the best for her progeny.
- Female fear of violence within and outside the family unit lead females to accept submissive roles and become the weaker sex by becoming more sexually appealing. Shame was attached to the menstrual cycle.
- Males developed an inflated sense of importance due to their role as protector and provider; protector in some cases becomes captor by restricting movements of the female from the home/house/hut, etc.
- Male bonding evolved into the early stages of the "old boy network"; males gained an understanding of the benefits of networking with similar thus "safe" males from one's specific hierarchy.

As we march forward in time from family units to the formation of larger social groups and small agrarian villages, we begin to see the protector role of the male more clearly emerge through defensive measures needed in order to protect their own villages against attacks from rival tribes. Obviously, coordinated defensive actions taken by males defending their villages are based on genetic self-preservation. Unfortunately we can also see the possibility that raiding parties would believe that mounting offensive attacks against peaceful villages that had more "stuff" than they possessed would result in a genetic advantage. This defensive posturing, added with the skills and knowledge of male-bonded hunting parties, contributed to their sense of importance and ranking within each tribe or village.

That is part of this new behavioral equation, but what appears to be new is that these organized raids by bonded males are the result of organized males following a charismatic alpha male. This alpha would have used his personal spiritual power to convince the group that physical violence was one of the best ways to become "king of the hill" in a larger territory. This larger territory would then become a larger pool of resources. What appears to be different is the emergence of one or a small set of elite males using beta males as a means of gathering resources for himself or this small set of elites. We have seen this in small raiding parties of chimpanzees making offensive moves against other primates. We can begin to see the darker side of man emerging (again) through the belief that such actions have a genetic advantage, only this time we can see the expanded cultural view that "might equals right," and that those who are weak are worthless and inferior. It can be deduced that male reasoning then led to an attitude that they obtain things women need and want for their progeny, and that women are only capable of staying home and tending children.

It is at this point that I want to discuss the "masculine" thought transference that males equate to other males who have failed individual tribal initiation rites. Studies throughout our species' early history have noted that group-centered, bonded males appear to have common initiation behavior mechanisms in place for accepting young males into their male-bonded ranks as adults. This involves young males passing certain tests of skill to "join their inner circle." Usually the feat has to do with overcoming some trial—be it through strength, bravery, or committing some act in a prescribed and customary manner. I believe these tests had been culturally constructed to equate the failing males to weak females, which would label them unacceptable. In early American-Indian tribes these males were ridiculed and made to stay behind with the women. The wording today in our culture of macho males disrespecting weak males refers to those males as effeminate (or worse) and has to do with the culturally constructed belief that weak males and weak females cannot defend their villages, homes, state, or country. But, of course, this house of cards is built upon cultural acceptance of the male being on top with the female as the auxiliary.

On our continuing voyage of discovery, this is the point in our early history where the success of the domestication of plants and animals and the invention of the plow also created the early stages of sociology—the science of the fundamental laws of social relations. Of course, we are at the most early stages of the relationships between family units evolving into farm villages, then into larger cities, and before the establishment of laws, courts, and institutions. But we have to acknowledge that we are at the point in our early history where we no longer take to the road in groups and decide as a group to remain in one general area.

What evolved are the raw, basic behaviors still planted in nature mixed with newfound decisions based on nurture to help us live in established groups. As such, we must also acknowledge that the early beginnings of social relations between individuals and groups in the jungle were originally based on physical domination by the stronger male. Recall the earlier world of the alphas and betas and their behaviors, which were the basis of the female's epiphany to leave that harsh world, devoid of choice. We are now at the point in our evolution where social hierarchies began to form. A hierarchy is simply any method of ranking one thing or person above another. In our particular case, it is Mother Nature's original process of sorting out in the animal world. This ranking mechanism evolved from the strong vs. the weak as the preferred selection process amongst animals, which dictated what genes would enable survival. But remember—that was then and this is now. Consider your world and your place in that world at a particular location and time in history.

This base, biological force that evolved from the individual struggle for survival was then carried over into grouped primates (and other socially connected animals), creating layers of social hierarchies. In all cases, in the upper layers of each hierarchy it has been confirmed that these individuals—humans and animal alike—enjoy more benefits than those in levels below them. In the lower animals, it would be called pecking order—a term coined in 1921 through the observation of dominant hens pecking at the heads of submissive chickens. In the higher animals—humans in particular—stored knowledge in their DNA (including the individualistic nature that aided our survival) compels them to resist submissiveness and strive to attain higher levels within their hierarchies.

Those in the lower levels of their hierarchies are aware of the benefits of living in the higher levels and want to be mobile enough and make all possible attempts to enter those higher ranks. This is also is true in reverse: those in higher levels know the negative conditions found in the lower reaches of society and do all they can to remain there or advance higher.

That was the wonder of America's success in the past: that with good public education, good opportunities, individual motivation, and encouragement and support from parents, one could become economically mobile and move up and into the higher ranks of social hierarchies.

Unfortunately this biological creation of hierarchies, which had its basis in Mother Nature's early plans of creating a better gene pool for survival, has evolved into the modern form of a conscious behavior based upon the balance of resources that one's family unit has managed to accumulate. One's social status is thus evaluated to be adequate or not in comparison with other family units and individuals. In America, one's position in a hierarchy has been reduced to a simplistic formula primarily associated with a monetary equivalent and the amount of "stuff" one accumulates.

CHAPTER 5

SPREADING THE SEEDS
OF ANDROCENTRISM

The Industrial Revolution

*During the Middle-Ages... The central work unit was
the marital couple, joined by their children when they
became old enough to work... Urban households often
included journeymen—but at their core in most parts of
Europe was a single marital couple and their children."*
(Wiesner-Hanks, et.al., 2011, p. 66 & 67)

If I took you on a whirlwind tour of history from the early
civilizations formed from the plow's agricultural successes
up to early rural England, stripped away all the kings and
queens and the nations created and destroyed, and only
viewed the passage of time without viewing the socially
constructed culture, what we find at its core are males,
females, and their relationships regarding the birth of progeny.

Since this is not a history book, what has been discussed
thus far shows how humanity slowly came to be androcentric. Obviously there are further ages of patriarchy between
where we left off with the invention of the plow and where
Darwin helped to codify the male bias against the female.

Suffice it to say that not much has changed across those ages.

Now the evolutionary questions we need to focus upon are—did the female select these series of biological events as the best course for her progeny? Or did the male use force to seize the biological advantage over the female? Or was it both?

* * *

In early England, large agricultural estates dominated the landscape, were owned by "landed gentry," and were not considered members of the peerage class. This hierarchy was created by British nobility to biologically identify themselves as separate from the "new rich." From the late 16th century to well into the mid-19th century, the gentry made themselves known as the class most closely involved in the law, military matters, and politics. Early English Parliament was composed mostly of this class, and political control in a certain local environment spanned over several generations. It is this layer of social hierarchy that I consider to be the basis for much of the behavioral social norms passed on to our Western culture, including the firm establishment of patriarchy.

> "European capitalism, including its gender patterns, grew out of the agricultural system that had developed in medieval Europe. During the Middle-Ages, the household became the basic unit of production in most parts of Europe, a process some social historians label the 'familialization of labor.' The central work unit was the marital couple, joined by their children when they became old enough to work… Urban households often included journeymen—but

at their core in most parts of Europe was a single marital couple and their children" (Wiesner-Hanks, et.al., 2011, p. 66 & 67).

In addition, society was flooded with new written physical evidence about male and female physiology through the invention of the printing press in 1440 CE. These were male-dominated publications of observations and private thoughts, and they helped rapidly spread and further reaffirm male-dominated culture.

As cities grew, the successful family farm needed materials other than just plows and seeds. Purchasing items that families had originally made eased everyone's condition. Items such as plates, utensils, pots and pans, clothing, wash tubs, horse shoes and feed, nails and building tools, furniture, and, of course, beer and a place to gather and drink it. All these items required someone to make them, and hence, this led to the rise of guilds, which were organized groups of craftsmen who supported each other and vouched for members' skills. And from the guilds we have the emergence of new occupations: arrowhead makers, bakers, brewers, bowyers, butchers, clothiers, weavers, cutlery makers, cooks, curriers, dyers, farriers, hat makers, sheermen, cobblers, tanners, and many more. But here we find also the firm establishment of male bias that excluded women from most guilds.

> "During the 13th and 14th centuries, urban producers of certain products began to form craft guilds in many cities to organize and regulate production; guilds set the rules by which most items were manufactured, including training requirements, quality and price levels, hours of operation, and size of the shops...in general the guilds were male organizations and followed the male life-cycle. One became an

apprentice at puberty, became a journeyman four to ten years later, traveled around learning from a number of masters, then settled down, married, opened one's own shop and worked at the same craft full-time until one died or got too old to work any longer" (Wiesner-Hanks, et al., 2011, p. 67A).

The male's freedom and need to move from one local environment to another to seek training and other resources furthered male patriarchy, deepened male bias, and laid the foundation for Darwin's flaw. We must acknowledge that one of the major pillars upholding patriarchy is the restriction of movement of the female partner by the male:

"This process presupposed that one would be free to travel, that on marriage one would acquire a wife as an assistant, and that pregnancy, childbirth, or child-rearing would never interfere with one's [the male's] labor. Transitions between these stages were marked by ceremonies, and master craftsmen were formally inscribed in guild registers and took part in governing the guild. Work continued to be carried out by a household unit, but only the male of that unit was recognized officially" (Wiesner-Hanks, et al., 2011, p. 67B).

So we have a family unit that has evolved beyond the agrarian norm of the Middle-Ages and is transforming because of the need for goods and services by a growing population. The need for more skilled workers and the need for more goods and services continued to increase and now we have a new development: the friction between masters and journeymen.

"By the late 15th century, journeymen began to resent the power of the masters...because women fit into guilds primarily through their relation with a master craftsman—as his wife, daughters, or servants—and their work was not recognized formally, they did not organize separately. By the 16th century journeymen resented even this informal participation, and increasingly asserted that the most honorable workplace was the one in which only men worked" (Wiesner-Hank, et al. 2011, p. 67C).

In addition to these new restrictions keeping women out of the guild workplace, the elite gentry in English Parliament passed a law to force young men and women to help them bring in the harvest on their estates and provide domestic help within the household:

> "The economy of this period was essentially a family and household one. We have seen how the institutions of service and apprenticeship brought a mass of adolescent and young men and women into households other than their own. There was virtually no place at all for the single man or woman taking up an occupation…Following the statute of artificers in 1563, mayors and justices could compel any unmarried woman between twelve and forty into service: adolescent girls who refused this course risked the insinuation they were whores" (Fletcher 1995, p. 228, Minchinton, ed, Nabbot, 1972).

> "Parody of female grievance by making a connection with insatiable female desire was in fact a common practice in the seventeenth century and the use of suggestive pseudonym names was a popular tactic… It seems men could only read disorderly female activity in sexual terms and thus constantly sought to nul-

lify women's taking up a public role by the sexual smear" (Fletcher, 1995, p. 8).

These two paragraphs indicate the social controls over women in Elizabethan England before the Victorian era. It is my speculation that in the bedrooms of the males who passed the Statute of Artificers in 1563 were females whispering into their ears and requesting assistance with household chores. Forcing people to provide all of this household labor would be framed by the larger view of acting on behalf of community well-being by providing the uneducated, strong, and healthy youth of the surrounding communities a skill and paying minimal wages.

The gist of the 1563 legislation is as follows: "Provided always, and be it enacted...that in the time of hay or corn harvest, the justices of peace, and also the constable or other head officer of every township, upon request and for the avoiding of the loss of any corn, grain, or hay, shall and may cause all such artificers [craft persons] and persons as be meet to labour ... to serve by the day for the mowing, reaping, shearing, getting, or inning of corn, grain, and hay, according to the skill and quality of the person; and that none of the said persons shall refuse so to do, upon pain to suffer imprisonments in the stocks by the space of two days and one night..." (Ditext.com).

> "The Statute of Artificers governs all trades and crafts. In the usual repetitive language of the law, it summarizes the acts and statutes relating to work and wages, vagrancy, apprenticeships, and price setting...It also governs retaining, departing, wages and obligations of apprentices, servants, and laborers: 'to banish idleness, advance husbandry, and yield unto the hired person a convenient proportion of wages.' No one may take up a trade or craft practiced in England

without serving at least seven years' apprenticeship" (Elizabethan.org, 2010).

It seems the original intent of the law was to reduce vagrancy and idleness amongst the young, and to fix wages. As we will explore the Victorian household of Darwin's home and his social ranking, the assistance given to the elite aristocratic female holds many clues as to the firm establishment of male bias intertwined with the agreement by their female mates, meaning that this was not just male bias against females but also one social hierarchy against another where they took advantage of the lower classes.

And speaking of the wives' point of view socially, how do you think one of the ways in which the female would present her plea to her aristocratic, Member of Parliament husband? This is mere speculation, but perhaps the elite female framed the argument to her Parliamentary husband that the more servants they had, the higher social status they could project to the outside world, and the higher per-ceived social ranking they projected, the more advantages they had socially.

We are backsliding into the biological pecking order and dominance displays, and I speculate that this time we have the female requesting help from her alpha male for the burden of raising children while increasing her social status—all to the benefit of her progeny.

If you read the entire law, it becomes clearer in the wording as to what this really means: that large family estate farms belonging to the landed-aristocracies have used their influence to pass laws to force laborers to stop what they are doing (in their line of craft-work in towns and small cit-ies) and help bring in the harvest in the their fields and to work in their homes doing domestic service work. Basically this is forced, cheap labor, but made legal by the passing of statutes and reinforced by justices of peace, constables, or "other head officers of every township." But what I find most appalling is the raw, vicious need to spread rumors by one social class as to the possible sexual behavior of young females if they refused to work under such a system, which seems to border on the simple delight in derogating someone from the lower classes.

So in 1653 we see constructed social behavior forcing young women to either submit to the law or run the risk of being accused as whores and thus destroying their chances of getting married and raising children in a worthy and moral marriage. What they were really saying was, "Young ladies, do as the law dictates or risk being banished to the taverns to work as prostitutes and be perpetually forced to remain a sexual slave in a male dominated society." Remember, rape is not about sex, but about domination and control. The domination and control are here, and the only thing absent is the apparent lack of choice in the sexual abuse she is subjected to. The law forced women to make the choice. And we all know that no well-connected, well-educated male aristocrat in early British society would ever take advantage of a poorly educated and non-socially connected chamber maid forced into service, right? And if the poor female did not submit to the sexual advances, what recourse did she have to report the abuse at that location and time in history?

This is an important moment in early British history and the spread of that culture of which the legislation entails. It is the "chiseling into stone," not just of male bias and male domination over women, but a new "legal justification for greed" viewpoint of the wealthy hierarchical classes placing restrictions on the movement of people of lower social ranking from one local environment to another, thus restricting their ability to seek their own level of opportunity. Or another way to view the law is that at a specific moment in time and location (all of England), we see the establishment of a codified legal document imposing restrictions on social behavior to economically benefit a few.

> "The household, we should remember, was a social construction not a natural one. It complemented the marriage patterns of the time and, as we have seen, provided a convenient focus for exacting discipline among the young. It suited the medieval system of guild regulations in towns. Indisputably it was shot through with sexual inequality, for, while the women brought resources to the family economy and worked hard to support it, control of that economy rested firmly in the hands of her husband. Whatever she did her housewifery defined her character whereas his work defined the objectives of family life…The universal assumptions were that women's labour was essential to the household economy and that the partnership in running it was an unequal one" (Fletcher, 1995, p. 229A).

While reading the following quotation, it's easy to envision Betty Friedan tapping away at her typewriter composing the *Feminine Mystique*. Try to remember that this is 1650 London: wipe away modern visions of the popular TV series, *Downtown Abbey* and think more of Bob Cratchit's lower middle-class family in the Charles Dickens' classic, *A Christmas Carol*.

> "There is no mystification in the frank account, which sounds strangely modern, of a housewife's day provided in the ballad 'A Woman's Work is Never Done,' published in 1650. While her husband

devoted himself single-mindedly to his occupation, she was required to be endlessly adaptable, tireless and patient. The day begins with sweeping and cleaning and making the fire. Children have to be dressed, given breakfast and got off to school. Cooking and preparing more food for a husband, who is quickly back at work with a 'scarce a kiss,' takes up the morning; knitting, spinning, washing and scouring the afternoon. Then children and husband need further attention. Nights are interrupted by a child crying for the breast and sex is about the only consolation" (Fletcher, 1995, p. 229B).

As populations grew, the continued spread of the printing press helped to speed up and spread the knowledge that one needed to navigate in this new social world. Assuredly this new culturally constructed knowledge would be spread the fastest by those who controlled the printing presses and the elites who made decisions as to what could be written and how far the information should be disseminated. And, of course, the spread of that culture would also depend on those who could read, comprehend, believe, and trust those words.

We have just left the Middle Ages, where there were basically just three social hierarchies: the nobility, the educated religious orders, and the rest of the population as dirt-poor farmers or peasants. Very likely this lower caste could not read. The plausible outcome would be that the illiterate would be in awe of those who could read and write and of the retelling of the knowledge found within this new technology. The first two layers of the hierarchy would be considered the elite leaders of society and the executors of what would be considered to be proper behavior within these narrow social confines. We see a society that is tightly controlled by the mysterious yet powerful religious order and their dictates. To reconfirm this history, remember Nicolaus Copernicus' theory of the Earth revolving around the Sun, how Galileo Galilei confirmed that theory via his telescope, and the Catholic Church's suppression of that knowledge for over 250 years. Why would the church do this? Always remember: a good gig with benefits is a hard thing to give up.

And we are now at an important moment in history when, because of the physical distance from Rome to London, we see the transformation of a dominant culture from its "birth of origin" overruled by another "holy" figure in London (as kings and queens were believed to have descended from God). The dominant cultural mass in England became aware of its territorial local environment, believed that the territory was sacrosanct, and overthrew another cultural rule (Rome). This major event took place because the "holy" King Henry the Eighth had a desire for a male heir. And since the Roman Catholic Church forbade divorce, King Henry dictated that his entire country's religion had to be changed so that he could have the opportunity to have that male heir. This is a major cultural shift and was not easily accepted at first by the lower social hierarchies.

The focus here is on the time period from 1500 to 1800 CE in England—and only England, not because we purposely want to ignore the rest of the planet's history, but because it is this fertile, male-dominated culturally constructed timeframe from which Darwin based his observations that the Victorian male of his day was "atop the evolutionary tree." By the time these patriarchal behaviors were repeated across generations, the knowledge passed from generation to generation had become cemented into cultural permanence as truth.

We start with a valuable lesson: that control of the culture of the "message" and how one person transmits those messages to another person are of primary importance. Some may even say this new navigation within ever expanding populations containing many layers of social hierarchies could be called *group selection*. Because if you do not know how to behave within the group you find yourself a part of, you will be intuitively ignored as an individual of inconsequence by the dominant culture. If you are viewed as a threat to the dominant culture because of what you say or do (or because you are the "wrong" race or religion), the possibility exists that you will not be allowed freedom to move from one hierarchy to another, thus restricting your choices. The ultimate expressional behavior by the dominant culture would be your elimination (e.g. the hanging of an African-American slave in the American Deep South by dominate whites). The significance of this cannot be underrated as the killing of your opponent, in alpha male terms, means you win (Allen, 2000).

The following quotation best sums up the cultural knowledge from 1500 to 1700: "The didactic [instructive] literature was written by men, some of it specifically to instruct women. So were ballads and plays…they tell us what women heard, saw, read or were taught. But they tell us nothing about what they thought…This is a reflection of the historical record: that this record is partial and that it provides a wholly inadequate basis for assessing women's private feelings and thoughts about early patriarchy… of men's understanding of the body, sexual activity, pregnancy and conception, the reader might well respond, if so minded, that women who knew about these things did not write and men who wrote did not know" (Fletcher, 1995, p. xxi).

I believe the most logical conclusion to draw from early views of gender and patriarchy is very simple: we should throw all the gender "knowledge" from the 16th century to the 20th century into the trash pile and look with fresh eyes from a perspective of the dance between biology and culture from all the new evidence being presented in the 21st century. If only men wrote about women's mental and physical states, and then they presented that knowledge to be totally biased to favor their gender and to justify their social hierarchical positioning—men on top, women on the bottom; men superior, women inferior. And the female, completely dominated into submissiveness, knew about their bodies but did not pass this knowledge on in any written form to be found by scientists, anthropologists, and philosophers.

Since we find a male-controlled economy from the plow to the guilds and religious orders dominating the cultural landscape, there had to be convincing evidence to continue those socially constructed male-restrictive behaviors over those considered inferior or below them for the "good of society."

> "There is extensive literature from this period showing an obsession with the dangers ascribed to womankind and inhabited by fear of women's motives and behaviour. Men's dilemmas were focused upon a stereotype of womankind which left them feeling intensely vulnerable and unprotected. Women were seen as possessing a lascivious, predatory and, most serious of all, once their desire was fully aroused, insatiable. In brief…an important element in male assumptions about women was they needed sex urgently following the menarche" (Fletcher, 1995, pp. 4 & 5).

Ah! The myth that the female uses sexual access to control men! So perhaps buried deep in religious dictates is the male's suspicion, based on word of mouth knowledge passed from male to male, that the female has a powerful behavioral strategy luring all men? Fact or wariness?

These obsessions are nothing more than myths that were passed from individuals to individuals in order to navigate within their social circles. So let me see if I get this straight: the taller, stronger, braver, and more aggressive male was afraid of the poor, undereducated, weak and inferior female, and in order to justify his dominance had to put further restrictions on her? Quite a reach of credulity, right?

"The drama [stage plays] of the period makes the most of the fact that the price of women's lasciviousness is often pregnancy and the punishment of painful childbirth that all Eve's daughters risk for the desire expressed in the story of the Fall" (Fletcher, 1995, p. 5). Note the reference to religion; hence the logical conclusion is that male bias is based on false biological knowledge intertwined with the reliance on male dominated culturally constructed religious dogma. Recall the passage from the prior chapter? Patriarchy was thus founded upon God's direction and women's natural physical inferiority?

The most important thought from the above statement is very clear: nowhere is there any mention of progeny—it's all about domination of the female by the male. No understanding, no compassion—just total domination and control. So if you were a male and observed all around you that these stereotypical beliefs of womankind were the prevailing masculine culture, it's clear that the slightest change in a woman's behavior, beyond total submissiveness, may indicate that the female could pose a threat to the male's control. After all, didn't the female object to sexual domination in the jungle by our aggressive male ancestors? But the difference between the jungle and the early days of English agrarian society is what the more modern female used before taking action: speech.

> "Men's control of women's speech...was at the heart of the early modern gender system...Speech represents personal agency. The woman who speaks neither in reply to a man nor in submissive request acts as an independent being who may well, it is assumed, end up with another man than her husband in her bed. Thus every incident of verbal assertiveness could awake the spectre of adultery and the dissolution of patriarchal order...Chaste, silent and obedient: the trilogy of primary female virtues carries with it a series of logical connections" (Fletcher, 1995, p. 12).

This parallels the thinking and attitudes in the deep American South in the 17th through the 20th century and the treatment of African-American slaves considered inferior by their white slave owners. The slightest sign of independent thinking on the part of the slave could have set off the belief that the slave wanted to flea his captor or, perhaps worse, the educated slave could organize other slaves in an act of defiance or revolt and inflict harm upon the slave owner's family. In this particular culture, at that general local environment, this "rebellion" by the slave could have ended with the slave being whipped mercilessly, the removal of part the slave's foot, and in the most extreme case, hanging from a tree by a noose—but this ultimate punishment would mean the loss of the slave's labor in the fields. After all, they were still considered functional and useful "property."

Now how could a female's free speech lead to having another male in her bed and lead to the destruction of the male patriarchal order? Of course the reasoning is illogical, but paranoia was behind the male act of restricting a woman's free speech as just another lock on another door to imprison and restrict the female's movements. The real reasoning behind male patriarchy is without a doubt to assure the male that the female he owns as "property" is not inseminated by another male, and that the mated male does not waste his time and energy raising some other male's progeny. I argue that behind this "imprisonment" behavior is the overarching, biological attempt by the male at avoiding cuckoldry.

I believe that this view by the male of the female's possible unfaithfulness is really a biological fear by the male that he could lose not just sexual access to his mated female by another male, but that she may desire another male over him and reject him outright. Jealously is a powerful emotion. We must always understand that strong, vital males possess a high sex drive, and that the mated male undoubtedly knows this of his fellow males as well.

The most logical method to prevent being cuckolded by other males was to ensure the isolation of the female and to restrict any means possible by which the female could stray. What this primarily boils down to is that the only way for the male to control this fear would be to maintain patriarchal order by physically restricting the female: first by verbal commands and then with threatening, elevated voice commands; and if that did not work, then via physical restraints and or violence. It all could all begin with the male's perceived disrespect shown him by the female through her speech and her tone.

"Men at this time were somewhat obsessed with a link between shrewish or scolding behavior and sexual infidelity…Proverbs which associated sound and sexuality hinted at the issue of female mastery which always had adultery as it ultimate and inherent point of direction" (Fletcher, 1995, p. 12 & 13). In other words, the most logical way for the male to avoid this behavior would be for males to close their ears to anything that a women said, felt, believed, or did because it would jeopardize the male patriarchal order of domination and control, which, men concluded, was behind all of humanity's successes up to this time in early pre-industrial England.

We have seen on this historical journey that the cultural "game was rigged" in favor of males since the widespread use of the plow, setting the stage in preparation for Darwin.

CHAPTER 6

MEN AMONGST MEN

Preparing for Darwin

*"The man who was not master in his own house courted
the scorn of his male associates, as well as economic ruin
and uncertain paternity."*
(Tosh, 1999, p. 3)

Since we are at the point in early English history that
would be considered pre-Darwinian and pre-Victorian, it
is important to look closer at the dominant culture of the
landed gentry who set or defined "gentlemanly" conduct in
early English history of patriarchy. The social norms that
flowed from those unwritten rules have drifted down to
us through the memes and institutions established in the
West. But take note: culture is fluid and never remains the
same.

> "For them [the gentry] honour in the Tudor and Stu-
> art periods presented a complex and demanding code
> of living and of behavior. Richard Brathwaite, himself
> from a Westmoreland landowning family, assumed
> in writing his book *The English Gentleman*, published
> in 1628, that a gentleman would wish his conduct to
> be guided by the dictates of honour…A gentleman's
> honour, in other words was the essence of his repu-

tation in the eyes of his social equals, providing him with his sense of worth and his claim to pride in his own community, contributing to his sense of identity with that community" (Fletcher, 1995, p. 126).

So now, with the masculine world fully established and with apparent approval of the female in their social hierarchies, so begins the further construction of building the outside of the house upon the foundation of patriarchy. Think of it more in terms of "male plumage" and a show of "stuff" to other males and family units—it is the essence of the beginning of man's view of himself in society in terms of his own masculinity. Once again, Anthony Fletcher:

> "Honour became for the gentry of Stuart England a complex and demanding code of living, expressed through notions of lineage, courage, virtue and reason in the various arenas of household, neighbourhood and local community. It was through seeking honour, it has been argued, that such men lived out their manhood" (Fletcher, 1995, p. 322).

It would appear that the higher one rose in the social hierarchies of early English society, the more "rules of behavior" one was expected to navigate within those hierarchies—polite manners, using the appropriate utensils at a formal dinner, knowing the steps to the dances of the season, etc. Think of it in terms of an "obstacle course of behavior": if you don't know how to jump over the obstacles, then you may not be suitable to maintain your social hierarchical position, and then you lose the potential for advancement into the next higher social order. I believe this is nothing more than one social group creating opportunities or obstacles for others to traverse in order to succeed within that social hierarchy—or to fail to enter it.

"The crucial new ingredient in English masculinity between 1660 and 1800 seems to be the notion of civility. This was more a matter of manners and outward behavior than of inner qualities, though for some it was seen to reflect these. The detailed attention being given by the conduct book literature of this period to a gentleman's dignity, argues Fenella Childs, 'reveals very clearly the power credited to manners as a vital means of expressing and upholding the social hierarchy.' Manners became the centerpiece of both the social and gender hierarchies, as masculinity in the upper ranks was more closely and deliberately defined and constructed than it had ever been previously" (Fletcher, 1995, p. 323).

These behavioral manners were a ticket for mobility into a higher social pecking order: "Social aspiration was a driving force in Hanoverian England, unifying the middle ranks in an almost obsessive engagement in seeking entry to the ranks of the gentry by aping their style of life, manners and morals...seeking status in terms of class, the middle section of society—rising tradesmen, merchants and professional men—was bound to develop a consciousness of its gendered forms...Politeness was always shot through with gender" (Fletcher, 1995, p. 325).

This voyage through the social pecking order of the landed gentry of early English history is meant to help us understand that the socially constructed world preceding Darwin's flaw was a monumental factor in his mate selection. Fletcher's book focuses on the explosion of educated leisure-class males spewing forth volumes of advice publications on how to behave within specific social hierarchies. Hence, they represent a sort of very early social media, giving social freedom to the reader—freedom to roam anywhere without restrictions—if you had the proper credentials and knew the "entrance rules."

Now this is important: there doesn't appear to be any advice on how to lower oneself into any lower social hierarchy. That would be silly, right? There were guided tours of London's slums in the mid-1800s, but that does not mean the upper reaches of that society wished to join the ranks of those lower tiers. But why not? It is a very similar question you might ask yourself as you exit a grocery store's checkout counter today in early springtime where you are assaulted by weight loss headlines on almost every magazine after the late December holidays. Why are we are not greeted with headlines that read, "Gain 15 Pounds Before Valentine's Day!!" The answers can be found rooted in biology and the importance of sexual attraction, which, of course, leads us to sexual activities that produce progeny. Can you see now how biology rules your everyday world?

Individually we are social creatures, and we all belong in a social hierarchy somewhere on the planet at this particular moment in time—unless, of course, someone has cloistered themselves. If we need to discover the social place of individuals or to be reminded of who we are in a particular local environment, all one has to do is to travel beyond that arbitrarily set territory and attempt to navigate a new environment by interacting with the established social elites. What would happen if one arrived without letters of introduction from other members of the social elite's hierarchy or if one behaved in a manner that did not follow the unwritten rules of "normal" behavior?

What are social norms, and how do they influence our behaviors? Norm is short for normal, as in the normal behavior that is acceptable at a particular location and time in history. These locations today can be very local, as in one block in a large metropolitan city, or they can be vast territorial areas where rural communities stretch many miles or kilometers. They affect how we think about others around us and how we are expected to behave.

Since we know that evolution is the adaption of a species to its local environment, which includes understanding and behaving and conforming to the social norms located there, a thorough understanding of the local environment and social norms of Darwin's time would give us a better understanding of how Darwin may have thought or behaved, and more importantly, which social norms may have affected his perception of the human male and female in Victorian England, thereby influencing his conclusions. To be quite frank, why did Darwin use science in his natural selection theory book, *On the Origin of Species by Means of Natural Selection*, but bend the science to mold to his culture in his *Descent of Man and Selection in Relation to Sex?*

To understand more about the cultural pressures surrounding his decision, we need to study the socially constructed world of Victorian England—in particular the social pecking order of males, females, and the three social hierarchical classes of his day: aristocracy, middle class, and the working classes. This final, lower-level stratum is the class that had to soil their hands with manual labor, but it also is the class that the upper elites could not dismiss so easily because it was the foundation upon which all social hierarchal structures perched.

We tend to believe that evolutionary pressures are merely forces acting upon individuals alone. But as humans we face a different set of evolutionary forces because we all live in some form of a social grouping, and most of the local environments in which we live include large concentrations of other humans such as hamlets, villages, or cities. Even though we may think we stand alone individually, we still must survive by interacting with other humans. We did not migrate out of Africa alone; we trekked the passage in groups and most likely in those groups there were leaders, followers, helpers, gatherers, hunters, nurturers, and protectors. The interactions and biological forces of group formation that helped us form successful, helpful, and nurturing groups are still with us today despite intensive efforts by some elites to perpetuate selfishness as manna from heaven.

There is no greater source of information concerning the culturally constructed, masculine world that existed just prior to, including, and immediately after Darwin's influential publications of *Origin* and his *Descent* book than John Tosh's highly detailed work, *A Man's Place: Masculinity and the Middle-Class Home in Victorian England*.

A Man's Place not only establishes beyond a doubt the patriarchal social influence that encircled Darwin, but also gives valuable insight concerning the power of scientific social organizations that controlled the scientific zeitgeist of England at the time *Origin* (1859) was published. These science groups were all male in their memberships, and the prevailing and consistent themes viewed the female gender—both physical and mental—as inferior to the male.

The main theme of *A Man's Place* defined masculinity in relationship to domesticity within the emerging middle-class of Victorian England. The book establishes the definition of middle-class in the Victorian Era by taking us back on the historical timeline to just before the industrial revolution—about 1815—and leads us up to the turn of the century—1901—the year of Queen Victoria's death. Victorian England provides a valuable link in studying a particular social society that evolved from an agrarian past to a modern industrial society that dominates Western thought (and some say, global thought) today. In English history, this is a period of dramatic social change within a very small historical timeline, and it provides an excellent microscopic view of the evolution of modern social behavior both good and bad.

Since the beginning of this trek starts on the farm, the book explores the male and female development on those establishments. Men and women worked together side by side in fields and in the home to secure the survival of the farm; they were linked in joint production. As farms grew because of technological advances in agriculture, the two-person farm took on helpers. At this stage, we see expanded households of room and board assistants living on the farm and in some cases actually living within the same household but in separate, yet attached quarters.

Here behaviorists can see the continuation of separate gender roles that flowed from the jungle, which were based on child care requirements necessitating the female to be close for nurturance with the male tending the fields and expanding the role of breadwinner and protector. Perhaps protector is the kind word. They were, in precise terms, *paterfamilias,* or patriarchal in their structure. In short, they were a behavioral mechanism that maintained domination and control over resources, including the female.

> "From the Reformation until the eighteenth century there was a vigorous advice literature on the aims and methods of domestic patriarchy…in its precise meaning of "father-rule," (the term evolved from Latin and the Romans) patriarchy remains an indispensable concept, not only because men have usually wielded authority within the home, but also because it has been necessary to their masculine self-respect that they do so. This was clearest with regard to the control of female sexuality.…if a husband wished to

be sure he was not providing for—or still worse passing on his property to—another man's child, then he must exercise surveillance over his wife's behaviour (sexuality)…The man who was not master in his own house courted the scorn of his male associates, as well as economic ruin and uncertain paternity" (Tosh, 1999, p. 3).

If the Victorian male was not "master of his ship," he could face economic ruin? Perhaps, then, one would think twice about stepping outside the bounds of social norms, such as publishing a theory that the inferior female selected her mating partner instead of the "highly evolved Victorian male." Was that Darwin's dilemma?

"For most of the nineteenth century home was widely held to be a man's place, not only in the sense of being his possession or fiefdom, but also as the place where his deepest needs were met…In an age when, in the estimation of the Victorians, economic and social advancement reached unprecedented levels, the men credited with these achievements were expected to be dutiful husbands and attentive fathers, devotees of hearth and family. The Victorians articulated an ideal of home against which men's conduct has been measured ever since" (Tosh, 1999, p. 1).

The following quote from Mr. Tosh seems to sum up one of the most important elements of the socially constructed world of the Victorian male household: "The domestic sphere then, is integral to masculinity. To establish a home, to protect it, to provide for it, to control it, and to train its young aspirants to manhood, have usually been essential to a man's good standing with his peers" (Tosh, 1999, p. 4).

The most important segments of that paragraph that we must focus on are "and to train its young aspirants to manhood" and "essential to a man's good standing with his peers." As can plainly be seen, there is no mention of the female in this equation. Can you almost peek into the past and visually conceive a group of bonded males heading out for the morning hunt while leaving the female behind to care for the children? Can you almost pretend to feel the emerging thought that only males are essential to the survival of the group?

> "The dominant belief in Victorian England was that women were not only inferior to men, but fundamentally different from them. They were not just a few notches lower on the scale of rationality or resolution, but set apart from the superior sex by natural endowment for specific tasks requiring distinctive attributes. Traditionally in Western society men had regarded women not as essentially different, but as less perfect version of themselves" (Tosh, 1999, p. 43).

Not much has changed since Aristotle's "great wisdom" about women. And what does the above bit of knowledge about male views of women have to do with survival of the fittest? The answer, of course, is nothing. It is simply the male gender, with his superior strength and aggressive tendencies creating a social climate in their modern world where evolved, "gentlemanly" males are in total control of social, economic, and political events—resources—and are allowed to create a cultural world that benefits mostly only their gender and some of the female gender accepting the submissive role. Why should it be any different? Hasn't history confirmed these social states of affairs stretching all the way back to great philosophers that the female is inferior?

And didn't the culture surrounding the Victorian household uphold the premise that only social, professional, and political connections between males were viable avenues to seek education, employment, and to conduct commerce? All of these behaviors were propped up and supported by the Victorian home.

Without good standing with his peers, the modern male could not make alliances that produced more resources for himself and his progeny. But let me emphasize again: this behavior must be classified within a specific timeline and location on the planet. If we study other family units in other locations and in different historical settings, we may find variations of what would be considered normal patriarchal behavior. This simple schema is a basic principle of our evolution from individuals to social beings identifying with an established group in which to function in our daily lives. Think of it this way: social norms were either accepted or rejected. If rejected, then one must find their own way, breaking from the family unit and starting one's own family unit. Which path would seem easier? Associate with the behavioral norm and adapt which produce alliances that would be beneficial, or reject those norms and attempt to find one's own way? Have the basic behavioral questions facing our species as to the direction in life we should travel changed much since 1859?

* * *

In the next developmental stage, we see the emergence of industrial-agrarian farms emerging, such as wool, flour, and food processing. It is within this developmental stage that we see the rise of this new middle-class. And since this new middle-class was not part of the landed aristocracy, they could not rely on those families or individuals who controlled Parliament to arrange and provide for their futures. Since personal property was still controlled and passed on to only the male of the family, professionally trained males were needed to continue to keep this pre-industrial machine moving along because of the resource wealth that it produced. In order to do that, the male elites of this new burgeoning class had to rely on providing higher forms of specialized education for their sons, which were not available at the local levels. Hence, this new demand spurred the wide-spread expansion of public and private institutions of higher learning in England—a whole new social phenomenon in and of itself. Along with the great expansion of public universities, we also see the emergence of the boarding school for young males—another social development. And, once again, no educational advancement for women was created.

The world became more complicated and sophisticated, and Darwin had to adapt to it or perish.

CHAPTER 7

THE DARWINIAN BUBBLE

Darwin's Zeitgeist

*"Let women be what God intended, a helpmate for man,
but with totally different duties and vocations."*
(Queen Victoria, 1870)

It's important at this point to look more closely at the definition of "middle class," and the general characteristic of the social climate of the times—the zeitgeist—that surrounded Darwin. Darwin had to adapt to his local environment and understand the social, political, and economic opportunities and roadblocks he faced. In order to succeed, he had to follow the social dictates within his local environment.

According to Tosh in *A Man's Place*, there were only three hierarchical classifications in Victorian England between 1830 and 1900: landowners, which were the aristocracy; the new professionals, which included scientists, doctors, lawyers, bureaucrats; and the new manufacturing class of business owners, who came to be known as "the Midland Class" or "bourgeois"; and finally, the proletariat, that layer of the population who got their hands dirty by using the sweat of their brows.

If you know your Medieval European history, not much has changed for several hundreds of years leading up to the Victorian Era and the Industrial Revolution except the lessening influence of royalty and the Roman Catholic Church on the daily lives of the majority populations.

In establishing this definition for the emerging middle-class, Tosh's *A Man's Place* gives us the little known knowledge about the "servant count" that assisted the rise of the middle class. With agrarian productivity improvements and land reforms in full swing, the need for excess farm populations who worked the land was vastly reduced. Millions of farm hands and assistants were forced off the land and in desperation had only one place to go: the cities. These massive migrations into the cities, primarily London, created a surplus pool of cheap labor, and from this labor pool arose one of the new professions for the laboring class: domestic helpers for the new rich. These were mostly females, who could help with the chores that once were strictly the farm wife's role. Men found employment as stable hands, carpenters, and gardeners. The need for forcing young people onto the farms to assist landed gentry via the 1563 arbitrator law vanished into the wind of non-usage.

The newly established middle class home was a display built primarily by and for the male—such as antlers on a large male buck. But now these family units and the structures that surrounded them became one entity in relation with other family units. "To the bourgeoisie the home was also the prime means of affirming social status—a medium of display intended to impress visitors and neighbours" [Tosh, 1999, p. 47].

> "The standard establishment for a securely based bourgeois family was three live-in servants: a cook, a housemaid and a nursemaid (or sometimes a parlour-maid). Wealthy households employed the full 'below stairs' complement of butler, footman, housekeeper, several maids, coachman, groom and gar-

dener. The employment of male servants was a mark of superior status since they usually cost more than female servants, and since the largest proportion did stable work, indicating that the master owned a horse and carriage. Male servants were deemed more difficult to manage, especially by the mistress, so the wife's undivided responsibility for domestic matters tended to intensify the preference for all-female staff" (Tosh, 1999, P. 19).

Now if the "standard establishment" classification for middle-class is three live-in servants, it's notable that "at any one time there were as many as a dozen servants employed in the Darwins' service, nine in residence in rooms mainly on the top floor of the house" (English Heritage, 1998). This clearly establishes Darwin's family as upper-middle class amongst Victorian hierarchies. Along with the rise of the middle-class and the use of servants to lift the burden from the female, we also see several cultural forces at work.

Religion became increasingly important in the household, as well as the female's role in relationship to that cultural force. Tosh discusses the rising influence of religious thought as an overlapping consideration on social behavior becoming justified into reality.

"Ultimately, the power of home rested on the twin authorities of nature and religion. The home was ordained by nature because its function and structure predated civil society and was the precondition for its reproduction. It carried the authority of religion because the family was the medium through which the divine purpose had worked in both the Old Testament and the New, and most of all in the life of Jesus Christ himself…At its most elevated, the idealizing of home extended to the belief that domestic virtues would triumph over a heartless world" (Tosh, 1999 p. 29).

The other cultural force at play within this new movement is the chafing between domesticity and masculinity caused by the open market system born of the industrial revolution. The origin of this conflict lies solely in the new development and unique behavior of the "commuting male." This development began solely because the necessity of wealth acquisition now depended on the male's need to commute to where the jobs within these new industries were located—in the inner cities and their close environs. As Tosh notes, "More importantly, as work became detached from home, so its association with a heartless commercial ethic became closer. Early Victorian social comment is full of the chasm between the morality of the home and the morality of business" (Tosh, 1999, p.30).

I am more inclined to look at the "big picture" of the evolutionary pressures on masculinity as "man in constant motion" against probable female demands that the male share in the responsibilities for his children, and that this shining moment in English history reflects a combination of events resulting in the best case scenario of domesticity for both genders in which to pass their genes.

But with males in control of businesses, science, politics, the media, and the ultimate influence in future generations—the flow of resource wealth—what chance did the female have of maintaining this domestic bliss? Despite the fact that Evangelical Christianity was at its peak in Victorian England, giving the female the moral high ground at home, the reality of the competitiveness of the industrial revolution, in combination with the *paterfamilias* responsibility of raising the male heir to be ready for the outside world, worked against the concept of domestic bliss. In a nutshell, this idea that *"the idealizing of home extended to the belief that domestic virtues would triumph over a heartless world"* (Tosh, 1999, p. 29) mentioned above never really had a chance.

Male dominance in all things began to dismiss the female's role in the upbringing of the young male as extremely detrimental in the long run because the outside world required him to develop different skills. Basically the thought developed that the female and the religious, moral world of the hearth had an effeminate effect on the young male, and that this would not work in the competitive and masculine "real" world. *A Man's Place* gives us multiple citations of men becoming irritated under the social requirement of domesticity and finding every excuse to not come home—pressures of work requirements, business dinners, association and social club meetings, and so forth were the usual excuses. This sentiment alone and the requirements of higher education and skills for young males created the demand for and establishment of boarding schools. There the young male heir could be educated in the real world of a competitive, masculine environment, and the young male would be removed from the powerful influence of the mother and the effeminizing effect of home.

We now turn to the final stages of this battle in the establishment of domesticity in Victorian England as we enter the latter stages of the 19th century and Darwin's appearance to influence the culture. "By the 1870s an astonishing growth in athleticism was under way in England. It included new sports like track athletics, rugby, hockey, tennis, badminton, and cycling; the expansion of existing sports like cricket, mountaineering and rowing, and one notable import, golf from Scotland....for many men sport held out the reassurance of an alternative way of life to the feminized home" (Tosh, 1999, p. 188).

Darwin's *The Descent of Man and Selection in Relation to Sex* was published in 1871 and coincided with this rise in masculine "athleticism." As an aside, this rise in athleticism in Victorian England seems to parallel the apparent rise in competitive athleticism amongst young males in America in the past 20 years, as well as the rise of the conservative political movement over the past 40 years. This coincides with the degradation of the economic foundation of the middle class while the upper 1% enjoys a golden era of prosperity. The economic parallels are almost spooky, but they do reconfirm the biological basis for the formation of social hierarchies.

So does this Victorian English athleticism seem to parallel America's military pre-emptive nature today (especially, when a "conservative" male is in the White House)? Because at the same time as this English movement aimed to take young males out of the domestic sphere and teach them to become competitive, the strength of the English Empire and the breadth of its territories was at its peak. The phrase "The Sun Never Sets on the British Empire" was on the tip of every Englishman's tongue. Pride in nationalistic accomplishments was prominent as the British flag was raised on every continent on the planet. For young men who were not born of privilege, the call of duty to serve the English crown in service to the Empire in faraway locales was a call from heaven to escape the misery of the inner cities; this call to military duty held out the prospects of a much more promising future.

Now, let's look at the final chapter of *A Man's Place* entitled: "The Flight from Domesticity." By the late 1870s, several generations of young people had emerged from beneath the strict moral thumb of Victorian England, and they wanted no more of it. Males in particular postponed marriage for as long as possible, giving rise to a whole population of "old-spinsters" in the respectable female populations. There was a greater dialogue in the culture about the drawbacks to marriage and domesticity, and it seems that the main disadvantage that emerged was "the check that it imposed on intimate relations between men" (Tosh, 1999, p. 172). No, Tosh was not suggesting intimate sexual contact between males, but instead the close bonds that have allied males together since our pre-humanoid days when they grouped together for common defensive and hunting excursions. This was part of man's innate nature and was now boosted by the rise of masculine adventure lifestyles made popular in the culture of their timeline. If you recall Chapter 1, you will remember that *Darwin's Flaw* was based on male "superior" intelligence, forged by skills acquired through hunting with their fellow males.

> "Quite suddenly in the mid-1880s a new genre of bestselling adventure fiction was born. For a generation, the most widely read novels had tended to deal with love and marriage, and thus to underwrite the claims of domesticity. A new group of writers headed by Robert Louis Stevenson and Henry Rider Haggard believed that the reading public had been starved of flesh-and-blood adventure...Men set off into the unknown, to fulfill their destiny unencumbered by feminine constraint or by emotional ties with home...Support and companionship are provided by the silent bonds of male friendship—what Kipling in an early novel called 'the austere love that springs up between men who have tugged at the same oar together...It is certainly true that from that point a sharp distinction grew up between men's and women's writing—sustained by Kipling, Conrad, and Conan Doyle himself" (Tosh, 1999, p. 174).

Evolutionary historians know full well that preceding the appearance of these masculine adventure books, a famous quote arose in the Victorian culture in Herbert Spencer's book, *Principles of Biology* in 1864: "This survival of the fittest, which I have here sought to express in mechanical terms, is that which Mr. Darwin has called 'natural selection,' or the preservation of favoured races in the struggle for life" (Spencer, 1864).

Did Spencer's words give permission from other Victorian celebrities to rally elite English male aristocrats to justify their imperialist march to civilize the world by planting their flags and their seeds of white male supremacy?

"We must find new lands from which we can easily obtain raw materials and at the same time exploit the cheap slave labor that is available from the natives of the colonies. The colonies would also provide a dumping ground for the surplus goods produced in our factories."

And: "Remember that you are an Englishman, and have consequently won first prize in the lottery of life" (Both quotes from Cecil Rhoades, adventurer and diamond entrepreneur, 1877) (Brainyguote.com, 1877A & B).

As I have written above, I see a parallel between Victorian England (including how their expansive masculine culture evolved) and America's rising athleticism, global business practices, and its own military pre-emptive expansionism that followed the 2011 terrorist attacks in New York City, Pennsylvania, and Washington D.C. We don't have the chilling social Darwinism of Herbert Spencer, or the adventure novels of Kipling, Stevenson, and Doyle dominating the culture, nor the lure of faraway locales to colonize, but there is the solid establishment of conservative talk-radio shows and hosts that have attracted millions of followers that fall into that category. The themes that the conservative radio hosts constantly blare from their mighty perches follow Victorian themes of "the-wealthy-know-best about creating wealth in the world," and they constantly warn about the "others" below their social ranks as lazy, unproductive, and a drain on the wallets of hard-working men and women. They share Spencer's negative views of "government handouts" to the poor because it only "encourages the survival of the 'unfittest.'"

Also, in the late 20th century, there has been a major expansion of masculine magazines labeled "laddie magazines" on America's newsstands. These magazines are generally devoted to "sports, sex, automobiles and alcohol" and are apparently an attempt to return "to the good 'ol days" of masculine superiority, now updated to include the Playboy Philosophy, wherein the Victoria's Secret lingerie-clad female is available to males for their sexual relief at a moment's notice. All of these magazines send the constant theme of competition, adventure, and, of course, advice on how to get ladies and perform during sex. The magazines usually contain numerous pictorial displays of the true theme of the magazine: the possibility of sexual access to the girl on the cover and those found within. Once again we see the full circle of evolution: the passing of genes into the next generation and the species seeking advice on how best to achieve this feat through masculine adventure, risk-taking, and competition. As our world becomes more complex, the problem of how to pass those genes becomes just as complex, but the innate need to pass our genes into the next generation remains the same.

A Man's Place is a goldmine of information devoted to masculine and feminine domesticity during a very important time in our human history. The legacy that is England has influenced much of what occurs in America today in terms of social hierarchies and how they function. And what occurs in America today has great influence on what occurs in the rest of the world.

Before Darwin's arrival, the rising middle class home was the supreme moral center of Victorian England with the male as provider and "captain" while the wife managed the household and nurtured the children. But the real emerging truth was that the Victorian Middle Class, as John Tosh tells us, was "shot full of contradictions" (Tosh, 1999, p. 47). The "moral center" was really a false front for appearances to others. The male found increasingly more pleasure in escaping the restrictions of the feminine centered home through men's clubs. And the so-called nurturing female, as guardian of home and hearth, relied more and more on help from governesses and nannies. As seen through these lenses, one has to wonder whether the wealthy male and female's tears of separation were genuine when the "little man" of the house was sent away to boarding school. Is there a biological "desire" to seek freedom from the "burden" of childcare?

> "By going to his club, the English gentleman could escape the domestic pressures of wife, children, and servants while surrounding himself with men of his intellectual and social rank. The club was, in effect, an exclusive and responsibility-free second home--a place to enjoy the power that was one of the special privileges of bearing the title of 'gentleman'" (NYU. edu).

> "Such clubs did much to set the social tone and were 'the cradle of sound public opinion in matters pertaining to manners, if not to morals'" (Rosen, 2010A).

"By the time of Victoria's ascension to the throne (1830), there were just over two dozen clubs in London and these still excluded all but noblemen, gentlemen, the services and the professional classes. To be a member of 'society' entailed being a member of at least one, and probably more, of the clubs. No person engaged in trade, from the lowest shopkeeper to the greatest merchant could hope for admission to these bastions of privilege and exclusivity. By the time of Queen Victoria's death, (in 1901) almost sixty-four years later, there were approximately one hundred and fifty clubs of which only seven had celebrated their centenary" (Rosen, 2010B).

Did Darwin belong to any men's clubs? I could not find any information concerning that fact. And since the clubs were an escape for the male, the farther the distance one had to travel to those clubs would have had an adverse effect on the enjoyment of the escape. Darwin's rural home in Downe was quite a distance from the train station, and he would have had to switch trains to get to London. Since Darwin preferred the quiet and isolation of his home, away from contact with strangers not of his ilk, Darwin's escape was intuitively his work and studies.

However, Thomas Huxley, considered to be Darwin's "Bull Dog" for his public stand in defense of the Natural Selection Theory, belonged to one of the most exclusive clubs in London, The Athenaeum Club. What better place to have an after dinner debate over brandy and cigars and influence the intellectual elite about Darwin's radical new theory?

But, of course, not all inhabitants of England sipped wine and smoked expensive cigars. There was a complete separation between the educated upper classes and those born without wealth, education, family connections, and the luck of being born male. When I recall the classic opening of the *Tale of Two Cities* by Charles Dickens, I now realize why Dickens' words reverberate throughout the decades with equal impact as Darwin's: "It was the best of times, it was the worst of times, it was the age of wisdom, it was the age of foolishness, it was the epoch of belief, it was the epoch of incredulity, it was the season of Light, it was the season of Darkness, it was the spring of hope, it was the winter of despair, we had everything before us, we had nothing before us, we were all going direct to Heaven, we were all going direct the other way" (Dickens, 1859).

Notice the publication date of *A Tale of Two Cities*—1859—the same year that Darwin published the *Origin of Species*. These are two observations by two men whose works have survived the ages—one scientific, the other cultural—published at the same time in history within approximate locations to each other (London). Born just three years apart—one born of wealth and privilege and the other born of poverty and struggle. One viewing the world from atop his lofty hierarchy and the other looking upward at the layers of social hierarchies above him. The lesson is that one's view of the local environment depends upon one's location within that environment. And although *A Tale of Two Cities* was about the overthrow of the aristocracy in Paris and France, it sent a shiver of angst amongst the English aristocracy that the same social upheaval could happen in their comfortable world just across the English Channel.

Recall that although London was the seat of power, England was dotted with large estates inhabited by wealthy agriculture-based aristocrats consisting of landed gentries who passed the laws in London. The word *aristocrat* comes from the Greek words *aristos* and *kratos*, which together literally means "ruled by the best." Royalty means that you were a direct descendent of the king or queen's families, while nobility was a title bestowed upon a person if they met certain requirements. Want to guess what those certain requirements might be?

How do you think it would affect a person growing up as a male knowing he was "one of the best?" Do you think he would have a gender and social bias in all he saw around him? In particular, how would he view gender roles knowing that his sisters and all women in England could not inherit land or money from their husbands or family because "that's just the way it was?" Would he grow up knowing that he was "different" from not just women but those young males in large cities sweeping chimneys or working in the new industrial factories? How would it affect his view of the inhabitants of societies found in far-away places on the planet? Would the young male think of himself as lucky or that "that's just the way things are?"

The nobility were granted their land by prior kings and queens for their loyalty to the crown. They maintained and passed on the ownerships of those lands through a tight grasp of economic control to their first-born males, a process called primogeniture. This "law of the land" that passes that wealth to the first-born male had a singular purpose.

This singular purpose was to keep the wealth and power the family had accumulated in one location instead of splintering the wealth into many smaller plots by way of the second- or third-born males. And since this biological and economic process only involved males, we now have the cultural basis for a solid male androcentric patriarchy with built-in financial and social support and incentives.

Experts on evolutionary psychology know that resources are the building blocks of evolution. Resources come in various elements. Several of the most important are wealth, family connections, educational background, location of home, and the separation from "low social values" (poverty). Of course, health is important as well, but those that I have just mentioned appear to be at the top of every ambitious family's agenda seeking higher ranking in one's social hierarchy.

I believe that now is an excellent time to show you something about Charles Darwin's place in his social hierarchy around the mid-19th century by sharing with you two opening paragraphs of Darwin's biography—one by the BBC, and one by an American website titled biography.com.

"Charles Robert Darwin was born on 12 February 1809 in Shrewsbury, Shropshire into a wealthy and well-connected family. His maternal grandfather was china manufacturer Josiah Wedgwood, while his paternal grandfather was Erasmus Darwin, one of the leading intellectuals of 18th century England" (BBC History).

"Naturalist Charles Robert Darwin was born on February 12, 1809, in the tiny merchant town of Shrewsbury, England. He was the second youngest of six children. Darwin came from a long line of scientists. His father, Dr. R.W. Darwin, was as a medical doctor, and his grandfather, Dr. Erasmus Darwin, was a renowned botanist. Darwin's mother, Susanna, died when he was only 8 years old. Darwin was a child of wealth and privilege who loved to explore nature" (biography.com).

Did you notice the difference? The biography presented by the "Official" British Government emphasizes Darwin's wealth and family connections, but the American biographical opening emphasized the scientific connections of his family. The reason that I wanted to present these two opening biographical paragraphs was to establish again the importance of wealth as a major ingredient in social hierarchies in early English history. The emphasis on accumulated wealth has now firmly replaced the biological element of our ancestral past in the jungle as the most important ingredient in being an "alpha" in our modern world—think "ruled by the best."

We are now ready to enter into Darwin's world; pay close attention as I rely heavily on Janet Browne's masterful biography of Darwin.

> "And on the other hand, he [Darwin] lived in a world in which heredity was an obvious organizing principle. The upper reaches of Victorian society were, after all, built on the notion of human pedigree and good breeding, not only in the sense that an individual's position in the existing social order depended to large degree on birth, but also in the heightened emphasis then laid on manners and the cultivation of taste and intellect...Darwin had every reason to muse on good and bad breeding among humans...Most members of his intellectual elite associated themselves with the rising ideologies of meritocracy, utilitarianism, and personal 'character,' a...sense of personal effort and determination under adversity, while for the most part enjoying inherited private incomes and status by birth. Darwin's position as a gentleman was secure" (Browne, 2002. pp. 277 & 278).

The obvious question, that someone like myself born in the lower social hierarchies might ask after reading Ms. Browne's statement would be—what adversity? And how much "personal effort and determination" was needed for

Darwin to acquire his next meal or a place to sleep for the night? When one wakes up in the mornings, they wake up in their local environment and have to plan their day within that specific environment and survive where they breathe, eat, sleep, mate, and provide for their progeny. If the rails are greased in one's favor with few or no obstacles in one's way, it makes it much easier to travel down the track to destinations or goals in life—and overcome "adversity."

Again, Ms. Browne:

> "Darwin certainly believed that the moral and cultural principles of his own people, and of his own day, were by far the highest that had emerged in evolutionary history. He believed that biology supported the marriage bond. He believed in innate male intellectual superiority, honed by the selective pressures of eons of hunting and fighting. The possibility of female choice among humans hardly ruffled the surface of his argument, although he repeatedly claimed that female choice was the primary motor for sexual selection in animals...Advanced human society, to Darwin's mind, was patriarchal, based on what we then assumed about primate behavior and the so-called 'natural' structure of civilized societies" (Browne, 2002. pg. 346).

Since Ms. Browne wrote that Darwin "believed...his own people...were by far the highest that had emerged in evolutionary history" (Browne, 2002, p. 345), I believe this is an excellent time to mention Darwin's cousin, Frances Galton. Mr. Galton, known as a polymath—or a person whose expertise spanned several fields of science—seemed to have been profoundly impacted by the publication of his cousin's 1859 book, *Origin*; in particular, the first chapter

"Variation under Domestication," concerning the breeding of domestic animals. Galton was also considered to be a "gentleman" who, like his cousin Darwin, did not have to labor with his hands and spent most of his life pursuing his own studies surrounded by the "adversity" of his own relative wealth.

Galton's main connection to Darwin was his attempt to measure "intelligence through hereditary," and in 1873 (three years after *Descent*), Galton asked his now famous cousin to list what mental characteristics Darwin knew that he possessed. Galton was attempting to quantify the "genius" of scientific men in Victorian England, which he believed depicted England's preeminence in the world (of course, with males at the helm). His 1874 book, *English Men of Science*, was based primarily on responses to a questionnaire that he sent out to members of the Royal Society of Science; and once again, the questionnaire went to males.

This implication, that intelligence was passed down through family members and that "breeding" could help to improve the "herd" would be called eugenics, a term that Galton himself coined. The historical trickle-down effect of this eugenics movement had a large following, particularly in California.

> "Eugenics was born as a scientific curiosity in the Victorian age. In 1863, Sir Francis Galton, a cousin of Charles Darwin, theorized that if talented people only married other talented people, the result would be measurably better offspring. At the turn of the last century, Galton's ideas were imported into the United States just as Gregor Mendel's principles of heredity were rediscovered. American eugenic advocates believed with religious fervor that the same

Mendelian concepts determining the color and size of peas, corn, and cattle also governed the social and intellectual character of man" (Black, 2003).

Now take this eugenics movement and attach it to Darwin's belief that the "Victorian male was atop the evolutionary tree," and the cultural effect would be like throwing gasoline on a Boy Scout campfire. It is even inferred that Adolph Hitler may have gleaned some knowledge from eugenics, thinking that if one were to eliminate the "inferior" members of society and encouraged "breeding" of the "superior" members, a nation could produce a "master race" that would rule the world.

This was the state of English culture and the intellectual climate (the Zeitgeist) permeating through Darwin's cultural timeline just prior to the peak of the British Empire. But take particular note here: there is no evidence that Darwin dipped into the racial or social implications of this cultural view that his fellow humans downwind of opportunity were inferior. But it did seem to enter his thoughts regarding women and birth control—I'll cover that in the conclusion.

While not quite at its peak during Darwin's rise to prominence, England was well on its upward path to becoming an empire that ruled the world and everyone in it. That's pretty heady stuff to someone who looks around within their local environment—filled with their fellow wealthy, educated countrymen—and knows in their bones that it couldn't be otherwise.

Why would they conclude otherwise?

CHAPTER 8

MOVING THE GOALPOSTS

Stopping the interloper

"Wallace found it difficult to get beyond the closed doors of London's most elite learned scientific societies. When he did, he felt out of place…it seems that he preferred the perils of the rain forest to the predatory jungle of metropolitan science"
(Browne, 2002, p. 28).

Do you recall in the prior chapter my thoughts describing the fictional relationship between Darwin and Dickens? "One viewing the world from atop his lofty hierarchy and the other looking upward at the layers of social hierarchies above him"? I included the quote from Dickens' *A Tale of Two Cities* to further inculcate your knowledge of the zeitgeist of Darwin's time period but also to introduce to you to how the social norms within these hierarchical layers would affect Darwin's publication of *Origin*.

Although there is no record of Dickens or Darwin meeting or corresponding, Darwin now had to face one of the most difficult events in his life: someone from the lower social ranks, like Dickens, quite literally almost knocking Darwin from his lofty perch amongst his scientific peers and destroying his primary work of the past twenty years.

It's my opinion that Darwin—isolated in rural England, surrounded by the serenity of nature and taking into consideration his lofty social hierarchy—was so comfortable and economically secure that there would have been little motivation to publish what he knew was a highly controversial theory that would tear at the entire social and religious fabric of Victorian England.

Given his serene life in the country—not to mention his psychosomatic stomach problems—without the series of events that were about to unfold, I speculate that Darwin had no desire to publish his *Origin of Species* until after his death—perhaps leaving that task to one of his eight remaining children, or more likely his strongest supporter, Thomas Huxley. Emma, Darwin's wife, was reportedly to have said to Darwin when he asked her if he should publish *Origin*: "Publish it or burn it." The mere fact that that Darwin worked and re-worked the natural selection theory for almost 20 years would suggest he was disinclined to publish.

* * *

In June of 1858, Darwin received a correspondence packet from a man named Alfred Russell Wallace that shocked Darwin to his core. To put it in evolutionary terms—Darwin was now faced with a bleak prospect: publish or perish.

> "What the packet enclosed was a short handwritten essay which, line by line, spelled out virtually the same theory of evolution by natural selection that Darwin believed was his alone…Darwin was stunned. [In correspondence, Darwin wrote] 'I never saw a more striking coincidence.' 'If Wallace had my MS sketch written out in 1842 he could not have

made a better abstract!'…These were probably the most lonely hours of his life, facing the knowledge that what mattered to him now was not so much the long-gone moment of discovery but the possession, the ownership, of his theory" (Brown, 2002, pg. 15).

Born of poor parents, Wallace was removed from school at age 14, so he never received a university education (yet, taught surveying and map making at a university). He wandered from job to job and finally found his purpose in the jungles of Malaysia collecting plant and animal specimens. For many years Wallace sat in the front row of that jungle observing and pondering nature as it changed before his very eyes. Wallace proposed that nature was the primary architect in a slow and gradual change to all species.

Darwin, having read every natural science article remotely linked to the subject up to that time, believed he was the only scientist that would have conceived of the natural selection theory. "Much of Darwin's shock apparently hinged on realising he had been thoroughly mistaken about the man. His years of dedicated research at Down House left him unprepared for seeing his hard-won insights reiterated, as he thought, by a nobody from nowhere"(Browne, 2002, p.23).

Ms. Browne's quote seems to contradict the citation from page 30 of her book *The Power of Place* that Darwin did read Wallace's article from the *Annals and Magazine of Natural Science* for 1855 in which he wrote, "Every species has come into existence coincident both in time and space with a pre-existing closely allied species " (Browne, 2002, p. 30).

Darwin did know Wallace somewhat because a year prior to Wallace's 1858 correspondence, Darwin asked Wallace via post to get him some skins of Malaysian poultry, and as such: "If he thought about Wallace at all, he probably would have regarded him merely as supplying basic data that he—Darwin—would turn into acceptable science…All he wanted from Wallace was exotic poultry skins and answers to a few specialised natural history questions" (Browne, 2002, p. 23).

Wallace was out of sight, out of mind to Darwin. Wallace did not travel in the same scientific circles as Darwin, nor did he have any connection with any of his colleagues; Wallace was not independently wealthy, nor part of any landed gentry families; he had no royal titles bestowed upon him and had no university-educated credentials. Simply put, Wallace was not even close to the same hierarchical level as Darwin. To Darwin, Wallace was "the other"—*"a nobody from nowhere"*—not in a mean-spirited or vicious way, but just invisible like a waiter in a restaurant or a postman delivering his mail, and thus, of little concern to Darwin. So Darwin had no reason to activate brain cells designed to signal caution or satisfy curiosity.

Darwin's concern with his possession and ownership of his theory is an important concept in the discussion of the sociological theory of social dominance, which we will cover soon. The meaning of possession and ownership imply something solid, tangible, and with constructed meaning attached as if it were important for survival, such as perhaps our early ancestors' strategy in capturing some four-footed prey, or knowing the location of a vast source of fruit trees that other early ancestors knew nothing about, or perhaps the knowledge of how to start a fire by rubbing two wooden sticks together.

* * *

Along with the correspondence packet that Darwin received from Wallace was an accompanying note in which Wallace writes: "...asked Darwin to pass the essay on to Sir Charles Lyell if it seemed sufficiently interesting... Since Lyell was often instrumental in bringing the work of unknown naturalists into the public eye, this was a reasonable request to make, and Wallace, who had no personal access to prominent scientific figures, manifestly hoped for the kind of friendly introduction that Darwin could provide...The essay in short, had been composed for Lyell, not for Darwin. Yet it would have been near-impossible at that period for Wallace to write directly to Lyell. A favourable word from Darwin would help him along" (Browne, 2002, pg. 15).

Sir Charles Lyell was a famous geologist and paleontologist whose major work, *Principles of Geology*, highly influenced the thought process of the young Darwin because it led him to think of the animal world and all its variations in much the same way that natural forces slowly acted upon the physical changes to the earth. *Principles of Geology* was the light bulb that lit up in Darwin's forehead concerning the variations of species._

> "Wallace found it difficult to get beyond the closed doors of London's most elite learned scientific societies. When he did, he felt out of place...it seems that he preferred the perils of the rain forest to the predatory jungle of metropolitan science" (Browne, 2002, p. 28).

But remember earlier when I quoted from Ms. Browne? "Darwin had every reason to muse on good and bad breeding among humans...Most members of his intellectual elite associated themselves with the rising ideologies of meritocracy, utilitarianism, and personal 'character,' a...sense

of personal effort and determination under adversity..."
(Browne, 2002. pp. 227 & 228). So, if Ms. Browne is correct about Darwin's social world, if someone worked hard and proved they were highly capable, would they not float to the top of that world like cream ascending atop coffee? Or were there unwritten rules attached to the concept of meritocracy that were a bit more complicated than simply stated? Was it because those who dwelled in the high social circles didn't want the "little people" to know about those pesky additional requirements for entry into their socially constructed world?

Wallace's self-image and our species' understanding of the invisible, "polite" rules of society governing a person's "passage" from a lower to higher hierarchy held back Wallace from contacting Lyell himself. This has all evolved from our primate past where violence determined hierarchical positions of alphas and betas—but we have "evolved" past violence; or have we?

Believing in these social norms of hierarchical structures as "solid doorways" that one could not enter unless ardently invited, Wallace wrote to Darwin and asked that Darwin pass the letter to Lyell because Wallace believed that Lyell, not Darwin, was more instrumental in "bringing the work of unknown naturalists into the public eye" (Browne, 2002, Pg. 15). How did Darwin, Lyell, and Hooker overcome this serious situation that the "lowly, socially positioned" Wallace could beat Darwin in the formal presentation of the natural selection theory? Those of us not in the academic world have to understand that whomever publishes first "owns" the right to stand on a hill, beat their chests, and shout to the world that they were the first to originate the idea. After all, once one achieves that "victory," the theory belongs to that person throughout history—this is not a trivial matter.

What series of events did Darwin and two of his colleagues perform in order to stop this serious situation of the "lowly" socially positioned Wallace beating Darwin—the one "atop the evolutionary tree"—to be first to posit the theory of natural selection?

* * *

Darwin was at a major crossroads in his life. He felt that his entire world and all the hard work he had put into the natural selection theory since his return from the Beagle voyage in 1836 was collapsing around him. On one hand he felt that he was driven by an invisible code of conduct of his social rank and that he should behave as an "honorable gentleman" and do the right thing. Yet on the other hand, in the highly competitive world of science, it would also be easy to have been tempted to remain secretive, since Darwin literally held Wallace's future in his hands with the paper manuscript Wallace sent to him; at last, Darwin's professionalism and honor won the day, at least momentarily.

A part of Darwin's dilemma knew far too well that to even suggest that his God did not create humans in some mystical garden of paradise or that man was the product of eons of adaptation would start a storm of controversy. Also he constantly read all the naturalists' journals and could find nothing even close to his theory. When he sketched out his theory to some of his associates, they did not seem to even glimpse the significance of the concept.

"As a consequence, Darwin felt that he was the only one who really understood his theory or could see how it might work as a biological explanation...It was hard to accept that he was not the innovator he imagined he was. Along with everything else, his scientific vanity was badly shaken" (Browne, 2002, p. 18).

Liberal thinkers in progressive circles of London for years were hinting that the advances of their industrial revolution era were the product of built-in advantages of their own breeding. Some of these thoughts were attributed to Darwin's grandfather, Dr. Erasmus Darwin, and to Wallace himself, who had published several articles during the 1850s hinting at the possibilities of distinctions between species and their varieties. So, with or without Darwin, scientific thinking was moving slowly toward a conclusion similar in form to his theory.

Resigned to his fate, Darwin then sent the manuscript as requested by Wallace and an accompanying letter to Lyell and in so many words confessed to Lyell that he was right—that he should have published when Lyell first suggested years ago, and that his words came true with a vengeance, and that all the originality of his theory was crushed to dust.

Lyell responded promptly and wrote in the strongest possible words that Darwin should publish some sort of statement along the same lines as soon as possible. But being a man of "honour," Darwin thought publishing his natural selection theory before the publication of Wallace's would shed light on him as a "paltry spirit." After receiving Lyell's letter recommending that he should publish some sort of statement, Darwin asked Lyell to discuss it with Joseph Hooker to get his opinion as well.

Joseph Dalton Hooker was one of the greatest botanists of early 19th century England. When Hooker returned from an expedition from Antarctica in 1843, Darwin approached him to classify the plants that he brought back from South America aboard the Beagle. Darwin included Hooker as one of his closest friends, as he'd confided to him his early ideas on natural selection. Their friendship lasted for close to 40 years, and the two exchanged around 1,400 letters between them—some of a highly personal nature touching on the deaths of their children. As an interesting aside, in 1851, Hooker married Frances Harriet Henslow, the daughter of Darwin's mentor, John Stevens Henslow, who originally arranged for Darwin to set sail on the Beagle and into the history books. This gives you a small window through which to view the close-knit circle of scientific and social friends that inhabited Darwin's world.

"Back came another letter a few days later. Lyell and Hooker suggested publishing Darwin and Wallace together, a compromise that accommodated everybody's needs as best as possible. With hindsight, it was the gentlemanly solution. Darwin's priority would not be lost, and he could carry on writing his big book; Wallace's views would be published in a way that would greatly enhance their interest and acceptability" (Browne, 2002, p. 34).

* * *

The next series of events for Lyell, Hooker, and Darwin was to decide how to accomplish this publication with both names attached to the submission. Lyell and Hooker decided they would publish at the next meeting of the Linnean Society of London on July 1, 1858. The last meeting of the year was already postponed due to the death of one of their former presidents, Robert Brown. Good luck, once again, shone upon Darwin with the gift of time—unfortunately, not so for Mr. Brown.

"They chose the Linnean Society for entirely opportunistic reasons. Lyell, Hooker, and Darwin were all fellows of the society and council members…Hooker virtually ran the journal and saw the programme secretary constantly…With these connections Lyell and Hooker could reasonably expect to have their way, much more so than if they had set their sights on the Royal Society of London" (Browne, 2002, p. 35).

Even Janet Browne, Darwin's biographer, had to bend over backwards for the proper word usage to describe this joint publication maneuver and could not sugarcoat the movement by the group that surrounded and protected the interests of Darwin, Lyell, and Hooker. Simply put, if you think you can't win the game, change the rules to favor your group's interests—and then move the goalposts without telling your opponent.

"Yet, for a while the proposal trembled on the edge of audacious skullduggery. No pair of practiced fixers could, if they wished, have cooked up a better scheme for promoting Darwin's interests. First and foremost, Wallace did not know anything about the proposal. His private communication to Darwin on a natural history matter, sent out to Lyell for comment, was to be announced without his knowledge and as an accompaniment to writings he knew nothing about" (Browne, 2002, p. 35).

"On the face of it, it looked as if Lyell and Hooker were suggesting that their friend Darwin—a man at the heart of the scientific society—should not lose out to an interloper—an unwanted visitor, trespasser, or meddler. On the broader scale, they may well have felt compelled to safeguard the values of elite Victorian knowledge—the science of accredited experts, authenticated fact, proper sequences of logical inference, and trustworthy sources—from outsiders like Wallace" (Browne, 2002, p. 35).

Do you see the pattern? There is an identifier phase the group uses to identify the outsider, and then they arrive at consensus as to what to do about the intruder. I would consider the group behavior by Darwin, Hooker, and Lyell as a mid-level form of group discrimination and contend that their decision to move forward as one entity was for the purpose of ensuring the safety and survival of that group's common interests. In evolutionary terms, the "action going forward" by the group would be about gathering and then protecting resources which assist in propelling their group's genes into the next generation over the "others." In the simplest language, Wallace was thrown under the bus to the advantage of a small group of elites. I wasn't there of course, but I strongly suggest that it was primarily the "audacious skullduggery" of Lyell and Hooker, with Darwin wincing in agony as a minor participant as the events unfolded. I liken Darwin's deference as having a tooth pulled at the dentist, knowing that a greater good will result after the painful experience passes.

When the printed material was presented at the Linnean Society, "Darwin's priority reverberated from every page. Even Darwin winched when he saw the layout some weeks later…He had assumed that his remarks would appear as a kind of appendix or as footnotes to Wallace. Privately embarrassed, he was relieved he had not supervised this printed reversal of fortunes" (Browne, 2002, p. 40).

Darwin had Hooker enclose a letter to Wallace about the events that unfolded and sent the letter off to him somewhere east of Singapore. The response took about six months and, to Darwin's relief, Wallace wrote back and accepted the series of events as actually beneficial to him and said he was honored by the joint publication.

> "Modestly, he [Wallace] accepted the lesser role of co-discoverer that was thrust upon him. Perhaps he realised there was little else he could do. What was done was done. By the time he knew about the dual announcement he was hardly in a position to make a fuss, and his innate manners probably told him to acquiesce graciously" (Browne, 2002, p. 44).

Wallace's attachment to Darwin's coattails were most likely more beneficial than if he had attempted to present his version of the natural selection theory to the world on his own. It also reemphasizes the power of social rank on everyday events, and that meritocracy is highly subjective to an individual's social position.

My own theory on this behavior by someone of the lower social hierarchies rejecting their current position and seeking acceptance from the upper social ranks is what I have labeled "Origin Denial, Resource Attainment Realignment Theory." In a modern political setting, it would be similar to a population voting for politicians and ballot issues that work against their own special needs but remaining convinced that somehow they would be more beneficial if they attach themselves to the policies of this individual politician or party. It seems that the need to be similar in thoughts and actions while aligned to a higher-ranked group would appear to be more beneficial than the actual benefits received. In sociology, this is called the social identity theory. The belief from the lower hierarchies is that perhaps if they behave like those on top and dress, speak, send their kids to similar schools, or live near those they admire, it will miraculously turn them into an alpha person or family, and they will reap the benefits accordingly.

The group behavior of Darwin and his two friends is assuredly innately understood, but to my knowledge it has never been identified properly—at least not until the social dominance theory was formally published in 1999 by two sociologists, Jim Sidanius and Felicia Pratto. And although the theory is deeply imbedded in the social arts, the authors do give a slight tip of the hat to evolution: "We attempt to…[present] a theory of group oppression that not only relies on thinking within contemporary social psychology, political sociology, and political science, but also include important ideas from evolutionary psychology" (Sidanius, 1999, p. 4).

With this book, *Darwin's Flaw*, I am here to strongly suggest that the big fish of our evolutionary origins is about to swallow the little fishes of the social sciences. One would not expect the dedicated sociologists to just surrender all their hard work and say to the world that they were shortsighted and could not see the bigger picture of our evolutionary origins. Once again, as I have written, a good gig with benefits is hard to give up. But we have to understand that social scientists, like the rest of us, evolved from the primates, and we all have our local priorities in which we struggle for existence and to give our children the best possible chance in life. Once again, evolutionary change is a marathon, not a sprint.

As an aside, I find it ironic that the social dominance theory was published by Cambridge University Press—the same university that today holds the Darwin Correspondence Project, a home to over 15,000 pieces of Darwin's correspondence.

What the social dominance theory accomplishes is to firmly attach citations and specific names to various group behaviors you have witnessed and may have always known existed around you for decades but which you could not articulate and categorize within a scientific context. Social dominance helps bring into clear focus such group behaviors such as race, class, gender, and age discriminations and the physical violence that often flows from these discriminations. The theory identifies the motivational bricks that form the biological and social foundations upon which all human societies are built—which, of course, all flow from our primate origins. The obvious benefit is that upon understanding the basic causes of all forms of social discriminations, policy makers can then make decisions that create positive social change for all.

The difficult part for politicians is finding the ovarios and testiclos to implement positive social changes.

And although the theory may seem light years away from Darwin, Lyell, and Hooker's "audacious skullduggery," which their hierarchy perpetrated against Wallace, it is my opinion that it is only separated by degrees. I would rate Darwin and his two amigos' discriminatory behavior against Wallace as a 5 or 6 on a scale of 10—with 10 being perhaps the beheading of a political prisoner and posting that beheading on the internet via social media.

Sidanius and Pratto open their Social Dominance Theory with a bold statement and ask the question of the ages:

"Despite tremendous effort and what appear to be our best efforts stretching over hundreds of years, discrimination, oppression, brutality, and tyranny remain all too common features of the human condition...we seem only to have increased the overall level of chaos, confusion, and intergroup truculence during the post-civil rights era and the resolution of the cold war. We see signs of this brutality all around us…Rather than resolving the problems of intergroup hostility, we merely appear to stumble from viciousness to viciousness. Why?" (Sidanius, 1999, p. 3).

* * *

Before we examine the all-encompassing social dominance theory, we should review some of the more prominent theories to date regarding group inequality, which will provide you with a much better understanding of gender inequality—particularly towards the female in mate choice selection—and the infinite-sided sphere of human behavior. This is a fascinating journey through some of the more interesting phases of our species' social behavior and the theories that have evolved and how we view ourselves:[1]

- **The Frustration-Aggression Hypothesis**—"… formulated by a group of Yale social scientists, the theory suggests that aggression towards less powerful people resulted from frustration at not achieving highly desired goals and that aggression toward more powerful people would be considered quite dangerous, (e.g. one's boss), and as a result, aggression would entail less danger of a resulting backlash if one's aggression would be directed toward groups of people below one's hierarchical standing" (p. 5).
- **Authoritarian Personality Theory (APT)**—"arguing under the banner that there is a personality syndrome labeled authoritarianism, social scientists in the 1950s argued that APT resulted from child-rearing practices that humiliated and deprecated the child (corporate punishment) and predicated parental affection on the child's immediate

---

1   Note: all of the underlined theories here are cited from [Sidanius, 1999]

and unquestioning obedience to the parents. This kind of subjugating environment was thought to predispose children toward thinking of human relations in terms of dominance and submission and to teach a particular orientation toward hierarchy; the vilification of those thought of as weak, humane, or deviant (e.g. ethnic minorities), and the glorification of those perceived to be strong and powerful (the western hero, the conquering warrior). As such, authoritarians were hypothesized to hold conservative economic and political views, and also be generally xenophobic, racist, and ethnocentric" (p. 6).

- **Psychological Uncertainty and Anxiety Models** —"…theory that fear of the uncertain creates preference for safe and conventional vocations, fear of death, and dislike of ambiguous art, when tends to enhance conservatism. Another theory stemming from this fear ethos is the terror management theory (TMT). TMT predicts that people find those with different cultural world views existentially threatening and are motivated either to assimilate their views, to convert them, or to derogate or even exterminate them, all in an effort to restore the cultural anxiety buffer" (p. 9).

- **Value and Value Conflict Theories**—Theories that emerge from attempting to understand people's attitudes and beliefs about politics, outgroups, and social policies by examining people's underlying values, "or the priorities given to basic principles of attitudes and beliefs that relate to both freedom and equality. Research has shown that the importance one attaches to freedom is unrelated to one's political leanings, although equality values are quite influential. Supporters of left-wing political parties and policies place much greater emphasis on the value of equality than do supporters of right-wing political parties. The value of equality has been found to be extremely important in determining attitudes toward specific policies (e.g. affirmative action)" (p. 10).

- **<u>Social-Cognitive Approach to Stereotyping</u>**—"The underlining theories emerging from stereotyping is that social stereotypes should first and foremost be seen as the result of basic and entirely normal information processing. In 1976, scientists theorized that people learn stereotypes because of a predisposition to receive associations among events. In particular, they reasoned that people perceive relatively unusual negative traits or behaviors and relatively unusual people, such as ethnic minorities as going together, resulting in negative group stereotypes. Since…social stigma increases psychological salience, this would then explain why these negative features and stigmatized social groups become associated in the mind." (p. 10).
- **<u>The Facile Activation of Social Stereotypes</u>**—"Based on evidence that people learn covariations very easily, and even unconsciously… [then]…once learned, social stereotypes are then quite easily and facilely activated" (p. 12).
- **<u>Stereotypes as Causal Explanations</u>**—Very similar to above, except driven by the need to explain and understand the behavior of others who belong to a variety of social groups. In 1972, scientists argued that when certain social groups disproportionately perform certain roles within the social system, people come to assume that all individuals within these groups have personal characteristics consistent with those roles (e.g., Mexican language speaking individuals in America washing dishes

or in car wash roles—my two suggestions, not the author's). This suggests that when people make internal attributes explain behavior attached to a role in society, they are likely to come up with a stereotype they already know as an explanation" (pp. 12 & 13).

- **The Contextual Sensitivity of Stereotypes**— In 1984, studies showed that "when people's outcomes were positively linked with those of a stranger, they paid more attention to individuating features of the person and relied less on stereotypes to form an impression that person." In 1993, "a scientist argued that people in positions of power, which is more typical of dominant group members, are unlikely to have to pay more attention to subordinates, and so are especially likely to stereotype. Thus, power inequalities are particularly likely to contribute to stereotyping." (p. 13).

- **The Tenacity and Self-Fulfilling Character of Social Stereotypes**—"Research has shown that stereotypes are often quite robust, tenacious, and long-lived. In fact, with the admission of tokens (e.g., women in a male-dominated profession) this tenacity persists because people are more likely to make internal rather than situational attributions for the actions of these tokens. Stereotypes not only can provoke self-confirming behavior in stereotyped others, but also can bias memory in ways that get people to recall stereotype-confirming "evidence" (p. 13 & 14).

* * *

Unlike the psychological models listed above, where the human focus is on the internal thought process (the *nature* or de-evolved state of human nature), the next section covered are the social-psychological behaviors of individuals as they begin to construct the *nurture* part of their world. These include the behaviors that develop as the individual tries to connect to, and become deeply embedded with, the absorption of cultural norms. Also influencing the individual is the strong desire to fit in and become accepted in the community.

- **Socialization and Social Leaning Theories**—"This approach of study assumes that the primary reason that individuals exhibit hostile, racist, and discriminatory behaviors toward others is because, from early childhood on, they have been socialized and trained to feel and behave this way" (p. 15).
- **Modern Racism Theories**—"Despite vast changes in attitudes toward blacks in America, most social scientists are in agreement that even though blatant and extreme forms of racism against African-Americans are now relegated to the past, more subtle and indirect forms of racism remain. In particular, research has now focused on the principle-implementation gap, or the apparent contradiction between White Americans' expressed support for the principle of racial equality and their consistent opposition to the implementation of any concrete policies that might actually promote racial equality in practice" (p. 16).
- **Realistic Group Conflict Theory**—"RGCT is the perception that one's group's gains is another's loss and translates into perceptions of possible group threat, which in turn causes prejudice against the outgroups, negative stereotyping of the outgroups, ingroup solidarity, awareness of ingroup identity, and internal cohesion, including intolerance of ingroup deviants, ethnocentrism, use of group

boundary markers, and discriminatory behavior" (p. 17).

- **Social Identity Theory**—The SIT theory argues "that humans have a general desire for positive social identity. When it is unclear what the meaning of the minimal group membership is, they construct the meaning to be positive so that it can reflect well on themselves. They do so by presuming their group's superiority and by allocating more to ingroups than to outgroups. This suggests that the more stable group boundaries are perceived to be, the more members of different groups will discriminate against each other. This argues that group conflict is likely to be minimized when both the superior and inferior groups accept the legitimacy of the elites rule, the less conflict there will be, and vice versa" (Sidanius, 1999, p. 19).

One of the most important points to consider with the SIT theory is that "low-status groups often acknowledge the superiority of high-status groups with respect to the high-status dimension and often discriminate in favor of high-status groups rather than in favor of their own low-status groups" (p. 20). Subordinates deny their origins as hopeless vehicles of advancement and attach themselves to the dominants in hopes of obtaining a better life for their children. As you recall above, I have called this behavior *Origin Denial, Resource Attainment Realignment Theory*. The behavior is so strong that some theorists believe that subordinates actually favor the company of dominants over their own ingroups.

* * *

As we approach the overarching social dominance theory, what we find is an attempt to connect the two major divisions in human behavior: nature and nurture; individual and group behavior. The first describes behaviors directly related to individual or inward thought conceptions (nature); the second are thoughts that result in behaviors of outward actions within groups (nurture). Or as Sidanius and Pratto put it in their academic wording: [the theory] "is neither strictly a psychological nor a sociological theory, but rather an attempt to connect the worlds of individual personality and attitudes with the domains of institutional behavior and social structure...The SDT begins with the basic observation that all human societies tend to be structured as systems of group-based social hierarchies, [and that] the dominant group is characterized by its possession of a disproportionately large share of positive social value, or all those material and symbolic things for which people strive" (Sidanius, 1999, p. 31).

Recall how, in Chapter 3, "The Plow," some families would have more surplus resources—stuff—than other families, and that the "others" may have been motivated by envy or greed to obtain those resources?

Sidanius and Pratto describe these surplus resources "for which people strive" as *Positive Social Value*, mentioned above, or in their interpretation: political authority and power; good and plentiful food (as in no food deserts); splendid homes; the best available heath care; wealth and high social status. The obvious opposite of these positive values would be *Negative Social Value*—low power and social status; high-risk and low-status occupations; relatively poor health care; poor food; modest or miserable homes; severe negative sanctions (e.g., prisons and death sentences) (Sidanius, 1999, p. 32).

After making the observation that all human social systems are group-based hierarchies, the authors then enter the most important stage of their theory: how the social dominance theory maintains itself through various behavioral mechanisms. Sidanius and Pratto then discuss group-based versus individual-based social hierarchies. In an individual-based hierarchy, the person enjoys high ranking within a hierarchy based solely on his or her own ability: (meritocracy), e.g. artistic, musical, political, athletic, mathematical, or scientific abilities. Although not mentioned in the book, one must reach the conclusion based on statistics alone that these people (like Oprah, Albert Einstein, Bill Gates, Andy Warhol) are extremely rare. If this is reality, then how do the vast majority of successful people get to the top of their respective hierarchies? The authors lead us to the obvious explanation by teaching us about group-based hierarchies in which one's position is based on individual abilities (meritocracy), plus the ascribed membership in groups based on race, religion, clan, tribe, lineage, linguistic/ethnic group, and personal social skills. Below is an excellent example of the power of place:

"For example, two children may both have the same level of native talent, individual drive, and personal ambition (Meritocracy.) However, if one child is of the upper class, has ambitious and well-connected parents, and attends the "right" schools, the chances are that the child will do quite well in life. On the other hand, for the other child growing up in an impoverished, dangerous, and sociogenic neighborhood and afflicted with inferior schools, chances are that the child will not do quite as well in life. This, of course, is simply to state the obvious. Even in modern, democratic, and multi-group societies, the achieved component of social status is, to a very significant degree, dependent on the social status and power of one's ascribed group membership" (Sidanius, 1999, p. 32 & 33).

The next factor to consider that revolves around the social dominance theory is how the system maintains and supports itself through *The Trimophic Sturucture of Group-Based Social Hierarchies.* Many sociologists agree that hierarchies consist of three distinct stratifications:

> "An age system—in which adults and middle-age people have disproportionate social power over children and younger adults.
> A gender system—in which males have disproportionate social and political power compared with females (patriarchy).
> Arbitrary-set system—which "is filled with socially constructed and highly salient groups based on characteristics such as clan, ethnicity, estate, nation, race, caste, social class, religious sect, regional grouping, or any other socially relevant group distinction that the human imagination is capable of constructing. In such systems, one group is materially and/or politically dominant over the other" (Sidanius, 1999, p. 33).

This last classification—arbitrary-set system—is huge. Just think about writing on the KKK alone, or about an entire social class conflict between religious groups. But since these are not my theories, and I am not a sociologist, I am not going to break these classifications down into smaller, identifiable groups.

And yet, this concept is very important, as it describes what some social commentators in America today would call "class warfare."

"Human social systems are subject to the counter-balancing influences of hierarchy-enhancing (HE) forces, producing and maintaining ever higher levels of group-based social inequality, and hierarchy-attenuating (HA) forces, producing greater levels of group-based social equality"(Sidanius, 1999, p. 38).

Let me rephrase the above statement in terms of Darwin's dilemma in regards to Alfred Wallace's situation: Darwin, Lyell, and Hooker did all they could to exclude Wallace through hierarchy-enhancing forces (not telling Wallace that he was going to be jointly published with Darwin), while Wallace attempted to apply an honest and self-motivated hierarchy-attenuating force (writing to Darwin to pass his natural selection theory on to Lyell) in an attempt to produce a greater level of social equality.

The social dominance theory is complicated and will continue to be so because of the various levels of socially constructed hierarchies found on our planet today. While complicated in a social sense, the behaviors identified really can be boiled down as simple human evolutionary behavior that evolved from the alphas and betas of our jungle origins: everyone strives to be on top and avoids being on the bottom. It's the various behaviors that the "top" layers engage in to keep themselves there and keep the "others" away from the obvious advantages that they enjoy that contributes to the complications. Those on top are able to discriminate in this manner because they have more resources available to them—money, connections, and policy enactments, i.e. control of police forces and local enforcement laws—and, of course, the highly motivating desire to stay there because of the obvious advantages.

Can you now understand why Darwin is buried inside Westminster Abby as a national treasure, placed there by groups of highly influential elites from highly ranked hierarchies, and that although he was jointly published as a co-authored equal of the natural selection theory, Alfred Russell Wallace is not? No one claims that life is fair.

Toward the end of this book we'll revisit the female of our species and her role in these hierarchy-attenuating (HA) forces and how the she has used them to her advantage within her own local socially constructed hierarchy vs. those females (the others) in hierarchies below her.

Sorry that I had to drag you through all of that social behavior business, but I felt it important for you to understand how Darwin's "meritocracy" was aided in part by his associates in "high places."

Now we can turn our focus on the mating dance in the 21st century—and, that should perk up your interest a bit.

CHAPTER 9

MATE CHOICE SCIENCE

Who really picks whom in the 21st century?

*Perhaps popular versions of the Bible should open with
the phrase, "In the beginning there was sex."*

Perhaps popular versions of the Bible should open with the
phrase, "In the beginning there was sex." Yes, my children,
you must come to that reality—you are the product of a
sexual coupling, and it is the first step in your personal
journey called life. We'll skip the sex coupling of your parents for the moment, but we will start your voyage beginning with your own personal physical Big Bang—born
of blood—an explosion of light—cold and bursting with
noise—both strange and loud. You have emerged into an
entirely new realm of strange stimulations, and you may
not be aware of it yet—but, you are frightened and you
don't know what to make of all the new and terrifying sen-
sations except cry out in desperation, and you are instinc-
tively drawn to the female that gave you birth for comfort
whether you are a male or female.

But before you were born, your parents had to find
one another and be sufficiently attracted to each other to
want to mate. They also had to send signals that they were
willing to mate, but because we live in a complicated world

far different from our naked ancestors, they had to navigate through all the obstacles and social rules our modern society puts in their path in order to preserve social order so that those who believe they rule the territory in which you dwell can remain calm in the face of change.

As you recall, in Darwin's social world the males in his environment had it much simpler than males do today. When we harken back to Darwin's social hierarchy, we are reminded of inherited wealth, comfort, and a home filled with servants who performed the childcare, including home-schooling, food preparation, house cleaning, and in the case of Darwin's servants, to provide transportation, carpentry, and garden upkeep. The females known to Darwin and his associates were considered adornments and auxiliaries of achievement, to be presented to the outside world as the glorious attainment of the male God living on Earth. The Victorian home, represented by the male head of the family, strived for the perfect attainment and examples of home, hearth, family, and faith. In raising children and conforming to the religious cultural environment, it was considered to be the perfect place to bring progeny into the world and one where the female was expected to remain confined, while the male roamed the outside world for provisions and provided protection.

In the perfect world created by males for males, in which Darwin and all males of his era benefited greatly, women were not allowed to inherit property, could not vote, could not enter schools of higher education—and perhaps most importantly, they could not publish their dreams, passions, frustrations and pain to the world. In the Zeitgeist of Darwin's era, women had little or no reproductive choices, as birth control methods were just beginning to rise above the horizon. I'll have a final word about Darwin's views on birth control in the Conclusion.

But that was then, and this is now. Let's set our dials on the time machine for, let's say, the 1970s or 1980s, as we wiz past females gaining the right to own property, to vote, to be allowed admittance into schools of higher education, and to publish their thoughts and grievances; they took the place of males on assembly lines when war beckoned males to the front, and the invention of the birth control pill gave female's reproductive choices that their female ancestors never could have had imagined.

We are no longer in Darwin's Victorian world of restrictions and expectations, but guess what? There still are roadblocks on the rocky road to love, and even today we still see outside influences restricting both the male and the female from having complete freedoms in choosing their mating partners.

In 1872, Darwin published *The Expression of the Emotions in Man and Animals*. It was pieced together from left over material he did not use in *Descent* and was one of the first publications to use photographs. In *Emotions*, Darwin attempted to link human emotions to their animal origins by primarily using examples of facial expressions. Perhaps Darwin thought the use of the photographs would be easier to convince the general public of the animal origins of emotions because photographs transcend language. I believe the huge controversy he stirred up between his stance on evolution and religious dogma made him think in unconventional methods about ways of explaining his position around the passionate and forceful attacks by religious intransigence.

In his usual, meticulous manner, Darwin detailed his observations of anger, anxiety, astonishment, blushing, dejection, despair, determination, devotion, disgust, fear, grief, guilt, hatred, helplessness, horror, "ill-temper," joy, "low and high-spirits," love, modesty, patience, pride, shame, sulkiness, surprise, and tender feelings. In all these observations, Darwin's main premise was to show that these visible signs of emotions expressed by humans in their facial and body movements were passed on to us from observable animal actions.

Darwin's *Emotions* was merely his fist peek into the complicated world of non-verbal communication—a stepping-stone in the direction of mate signals between our modern sisters and brothers on their way to produce progeny. There is nothing more subtle and fascinating then the mating signals both modern females and males use to reach their reproductive destination.

* * *

Now, at this moment, recall Chapter 2 and my theory of how the female used her sexuality to tempt the beta males to exit the jungle. Also recall that language was plausibly not a form of communication, and since our pre-humanoid ancestors had facial hair covering the muscles beneath their facial skin, we can reasonably assume that body language was a more significant form of communication between the male and female. After all, the primary signals our early ancestors used to judge each other's mating intentions were the female's estrous sac that became robustly engorged and a glorious pink color and would be positioned in such a manner as to give our male ancestors a marvelous view. Obviously the female could view her own attempt at coupling with a suitable male if he responded with a healthy erect penis.

We will get into body language in more detail shortly, but let's ask an important question: why have sex at all?

The simple answer is that diversity makes us stronger. Mother Nature, in all her wisdom, knew that if we were capable of reproducing asexually, then the building blocks of which we are constructed would also be replicated. In typical science-fiction end-of-the-world-scenarios, if a pandemic were to strike us and our self-replicated genes were not capable of fending off the virulent strain of the disease, our entire population could be wiped out.

But if we were to mix our genes with others that have different building blocks of genes and the same pandemic would strike, there would be a greater chance that some of our ancestors would survive. So remember, it's not about survival of the fittest, strongest, fastest, bravest, etc., but instead about who leaves behind the most progeny—the true test of evolution.

Most people understand that their genetic makeup is made of half of their mother's genes and half of their father's. So, keeping the above scenario of a global pandemic in mind and the mixing of diverse genetic makeup, it would stand to reason that the greater the dissimilarity between the genes of mating partners, the stronger the genetic defense against diseases their progeny would have.

But how do we reach this perfect mixture? For this, our bodies have an immune system, and we resist germs genetically through the major histocompatibility complex (MHC). The MHC genes produce molecules that equip the immune system with the ability to identify harmful pathogens. Humans are genetically predisposed to choose mates with dissimilar genes.

There is a fun-filled science experiment that has exploded in popularity in college and high-school classrooms over the past several years designed to demonstrate Mother Nature's wisdom via the MHC: the smelly-T shirt experiment.

It's a simple hands-on (nay, underarm) experiment that is very easy to replicate and is a great tool that goes right to the core of genetic diversity. In a nutshell, brand new t-shirts were distributed to males in a classroom with instructions to wear the T-shirts non-stop for a 48-hour period and not to shower nor use any deodorant or body scent sprays. When the males completed that task, they put the T-shirts into sealed zip-lock plastic bags and returned them to the classroom where the clear plastic baggies were labeled alphabetically or whatever way "the boy-in-the-bag" would be anonymous to the female. It obviously works best when it is the female of our species that does the smelling because it is through her body that we reproduce new members of our species.

The females in the classroom were then asked to remove the t-shirts from the baggies and take a good, long whiff, focusing on the underarm area. The results were always unanimous: the females who were mid-cycle, near their time of ovulation, found that the most sexually appealing smell came from a male whose MHC was the most dissimilar from hers. If there would be any successful sexual bonding between the female and the male who captured her heart's (nose's?) desire, it would result in progeny with genetic building blocks that had a higher resistance to disease.

To confirm the validity of the demonstration, those females who were on contraceptives, which basically fools the female brain into thinking that she is already pregnant, the results were the opposite: they would not prefer to mate with that particular male. (Wedekind, et al. 1995, pp. 245-249).

"A PET-scan study by Ivanka Savic and his colleagues at Stockholm's Huddinge University Hospital found that estrogen-like compounds affect sexually respon-sive parts of a man's hypothalamus (the paraventricu-lar and dorsomedial nuclei) but not those of a woman. Vice Versa, testosterone-like substances stimulate sexually responsive parts of a woman's hypothalamus (the preoptic and ventromedial nuclei) but not those of a man's...A woman is most sensitive to a man's apocrine at the midpoint of her menstrual cycle as ovulation takes place " (Givens, 2005, pp. 192 & 3).

While we are on the subject of smell, there is wide doc-umentation that many animal species can communicate through chemical signals called pheromones. These are chemical messages that are released by the animal and can alter the animal's reproductive and sexual behavior.

In a recent issue of the journal *Neuron*, studies found that "human pheromones do exist, and that women can communicate with men and vice versa. This is a very important finding because it shows specific areas of the brain that are activated by these chemicals" (Malick, 2015).

However, I do want to express caution that most of our fellow humans have hygienic habits that cleanse those pri-mary areas responsible for the production of pheromones, i.e, underarms and genital areas, and do so on a frequent basis reducing any effective pheromone creation and com-munication (recall our t-shirt experiment and the rule that males do not bathe for 48 hours?). Along with frequent cleansing, this includes using scented cleansing materials, perfumes, and aftershaves that send their own messages. In recent years, the popularity of completely removing pubic hair from both sex's genitals for that "new, clean, sexy look" has been added to the equation of further deterring phero-mone communication.

Which means, of course, that Darwin had no clue when he penned the *Descent* that animals use pheromones to communicate sexual compatibility. Something else was happening beneath the surface besides the peahen's aesthetic preference for the peacock's flowery plumage as a gauge of genetic superiority. However, the weak influence of pheromones on the human mating sequence in Darwin's timeline, and especially today, just continues to add evidence that we humans have greatly reduced one "natural" element in our selection process of a suitable mate and now rely on additional input from the outside physical world in which we dwell—"nurture."

But we haven't completely left behind our animal connections. You will recall earlier in the book that I stressed body language as the primary method of communicating desirability to mate between our ancestors in order that Mother Nature's disease prevention mechanism, the MHC, could be put in place. But that was when our ancestors were still in the jungle, naked and free to "express themselves, 'naturally.'" But now we are all fully clothed, speak a complicated language, live in family units that are located in specific locations on the planet, and surrounded by rules of social norms created at a specific time in history that we did not put into place but have to learn to navigate in order to get along. We have to find our own way in the world, provide for ourselves, and then, if we sense the chemical "urges" moving us in the direction required and all the socio-economic ducks are in a row, we will ultimately find a mate and reproduce.

It stands to reason that, in our complicated world, getting our males and females to the final stage of gene exchange in order to reproduce is more complicated as well. It is at this point where we leave Darwin and his one flaw in the dust and turn to modern research.

In our timeline, so much material has been written about sexual attraction and the mating dance that it really is difficult to determine which stands out as the primary explanations. In this book, I will only dip into the works of four resources—two of them heavily: David Givens, for his book, *Love Signals: A Practical Filed Guide to the Body;* Timothy Perper, for his seminal work, *Sex Signals: The Biology of Love;* Ellen Fien and Sherrie Schneider, *The Rules: Time Tested Secrets for Capturing the Heart of Mr. Right,* and Helen Fisher from the Journal Human Nature: *Lust, Attraction, and Attachment in Mammalian Reproduction.* With the exceptions of *The Rules,* the three other sources have taught us that the mating sequence falls into attraction, escalating physical contact and finally the physical transfer of genes between the two sexes. What differs is the location and cultural norms affecting the two participants in a specific timeline in history. I want to add that there will be more from Dr. Fisher in the final section of this book as we speculate on "what if" gender roles were reversed.

I think Timothy Perper, in his book, *Sex Signals: The Biology of Love,* most aptly framed the frustration potential modern lovers face in order to produce progeny when he wrote profusely, although, a bit flowery:

"Everything, it is said, is against our lovers—sexual repressions that reach back into childhood, male-dominated rules that repress women sexually, an indwelling and built-in natural fear of pregnancy and intercourse, sexual moralities that set forth iron-clad Thou Shalt Not until marriage, terrors and dangers of social ruination, shame, and disgrace, thunderings from pulpits both real and televised against the scourge of promiscuity, sex educators who teach responsibility in making reproductive decisions. Clearly, only the most extraordinarily and profoundly dysfunctional and alienated sociopaths could possibly ignore such powerful injunctions, pressures, and rules!...What precisely, could...overthrow the immense forces of morality, education, religion, fear, common sense, and sexual repressions that date from childhood?" (Perper, 1985, pp. 125 & 126).

Perper's approach is deeply academic and philosophical and goes into minute detail about his theories surrounding the perceptivity of the female, and "the woman's role in initiating and maintaining the courtship sequences itself" (Perper, 1985, pp. 125 & 126). The major error that I see in Perper's presentation is the small sample and the narrow cultural location of those observations. He gleaned the majority of his study on essays written by only 117 female subjects on how they approach males and ultimately become intimate. He does detail his work in the concluding appendix by stating that more than 900 hours of actual observations in disco bars were done. However, my skepticism seems to beg the question: would Perper have found the same aggressive mating sequences by females he found in disco bars in the late 1970s surrounding a small college town on the East Coast of America in comparison to a small neighborhood bar in Topeka, Kansas in 1950?

In David Given's *Love Signals*, he breaks the courting sequence down into five phases, and first and foremost is :

> "Attracting Attention, [where] you advertise your physical presence, your gender, and your willingness to be approached. In Phase Two, Recognition Phase, you read how others respond to your bids for attention. Positive feedback invites you ahead to Phase Three: Speech. As you speak, nonverbal messages go back and forth inviting you closer—if all goes well—to Phase Four. In the fourth or Touch Phase, you transcend the logic of words and communicate

in a more ancient and more tactile mode. Finally, if courtship is successful, you validate your sexual bond in the final Phase: Lovemaking" (Givens, 2005, pp. XXI & XVII).

In the beginning there is you and your potential mate, and the first thing that you want to do is to send the message that you mean no harm. "In the first or Attention Phase well before courtship, people beam out signals to announce 'I am here' and 'I am female' (or 'male'). With their clothing, facial adornment, aromas, gestures, and deeds…nonverbal messages are broadcast in all directions to attract notice, well before words are exchanged" (Givens, 2005, p. 16).

Phase Two of Given's field guide on how to read a gleam in the eyes is also the:

> "Recognition Phase… [which] begins as you seek nonverbal response to signs emitted in the Attention Phase…Recognition cues give information about having been noticed. They are incoming signs received in response to outgoing cues previously sent…Recognition cues show where you stand in a relationship before you say hello" (Givens, 2005, p. 17).

The next step, Phase Three, is the:

> "Conversation Phase, [where] signs exchanged in the previous phases enable couples to penetrate the unseen barrier of stranger anxiety…Some think you shouldn't talk to strangers without having something witty or important to say. Remember that courting is 99 percent nonverbal. What is said matters less than the saying…Of course, social psychologists who do the research don't consider the preparatory gestures

needed to spark a conversation in the first place... how lip, eye, brow, face, head, shoulder, arm, hand, and finger movements help or hinder your spoken remarks. At the close quarters of speaking face-to-face, nonverbal signs of liking, trust, deceit, and willingness to commit are available for the reading" (Givens, 2005, p. 18).

Language of Touch is Givens' Phase Four:

"The Touching Phase begins with the first tactile contact, from an "accidental" knee-brushing beneath a table to a more deliberate tap on the shoulder or back. After smell, touch is humankind's oldest sense. So powerful are touch cues that initial body contact must be made with care" (Givens, 2005, p. 19).

Of course, *Love Signals* ends its courtship phases with the fifth and final of them: Making Love.

"The most intimate stage of courtship is, like the phases before it, replete with nonverbal cues. Embraces, pats, en face gazes, snuggles, nuzzles, cuddles, and kisses prevail as couples care for, handle, treat each other tenderly as babies...Couples exchange words in softer, higher-pitched voices. Physically through sound, words caress as gently and persuasively as fingertips...Our brain still responds to love talk as an intimate form of 'touching'" (Givens, 2005, p. 20).

Givens concludes his introductory chapter with a very important word to us all: "Nonverbal signals rouse deeper parts of the emotional brain, where mating instincts lie" (Givens, 2005, p. 21).

In his *Love Signals*, Givens attempts to give us a balanced view of the mating sequence between the two genders, but Perper, in *Sex Signals,* posits the view that it is the female who initiates the mating strategy and then allows the male to "take over" the mating sequence in what he calls the initiative transfer. When Perper did a post-question analysis of the mating sequence in his sampling, he discovered that only 1 out of 31 males could remember the entire mating sequence from who initiated the conversation, what was discussed, and who touched whom first, but *all* of the men remembered their "job" of what they did to excite the female sexually to the successful implantation of their seed.

The overall cultural view in male-dominated Western societies is that men are the seekers and seducers, while the female is the shy, coy, and submissive receiver of the overtures. Helen Fisher believes—and I concurred here in chapter 3—that this belief is a holdover from our agricultural past when young, virginal female daughters were considered property that was worthless around a plow yet highly desirable by bachelor males seeking long-term sexual access (Fisher, 1992, p. 279). This early agricultural period was the trailhead path towards our temporary patriarchal dominated culture. And, once again, by 'temporary' I mean the time distance travels on the Big Bang Cosmic Calendar Clock: a few seconds—where one second is equal to 500 years.

Western cultural beliefs have a long history of confusing myths regarding female sexuality: from coy and submissive, to dedicated mother and nurturer, to the sexual predator vamp who prowls the landscape with an instable sexual appetite, scaring the bejesus out of men who fear loss of control, primarily so that they would not be scorned by their fellow males.

In Fein and Schneider's book, *The Rules*, the authors discuss how rules were passed on via word of mouth. Do you recall the discussion in an earlier chapter about the myths of the menstrual cycle in India? Same difference. Despite the lack of "scientific evidence," the passage of information via word of mouth is the most trusted method of cultural exchange between family members and friends. The optimal reasoning behind the information exchanged? Confidence and trust. There may be a hard-wired space in the brain devoted to trusting information, but it does stand to reason that from an evolutionary stand point, we listen and trust information from family and friends more fervently than from educational material through other sources.

> "No one seems to remember exactly how *The Rules* got started, but we think they began circa 1917 with Melanie's grandmother…Grandma passed on her know-how to Melanie's mother, who passed it on to Melaine…But when Melanie got married in 1981, she freely offered this old-fashioned advice to her single college friends and coworkers, like us" (Fein, 1995, pg. 1).

I surmise that the overall meme of the advice to other females on how to "capture Mr. 'Right'" was for them to be "passive-aggressive." *The Rules* does not discuss the court-ship sequence nor body language and goes straight to the underlining stratagems needed to take advantage of what they believe males seek out females for in the first place: sexual access. You almost get the sense that the authors treat sexual access as a precious commodity, and that the scarcer the commodity becomes, the more the consumer– in this case, males seeking females—would be willing to "pay" for that access. So this begs the question in my mind: are *The Rules* just for single females who are considered attractive, sexually desirous, and highly sought after in our aesthetically driven culture?

We must remember that the sexual urge in some males is so strong that police blotters are full of men being arrested for behavior associated with the efforts associated with sexual relief. Some include public indecencies such as exposing their genitals to females in public, driving naked waist down in their cars looking for someone to have sex with, having sex with trees, and of course, forcing females to provide sexual access without their permission—date rape and violent rape. It is this aggressive urge, and the female desire to shape or mold that aggressive urge within cultural social norms, that appears to be at the heart of *The Rules* in its attempt at advice for women.

Rule 1. Be a "Creature Unlike Any Other."
Rule 2. Don't Talk to a Man First (and Don't Ask Him to Dance).
Rule 3. Don't Stare at Men or Talk Too Much.
Rule 4 Don't Meet Him Halfway or Go Dutch on a Date.
Rule 5 Don't Call Him and Rarely Return His Calls.
Rule 6 Always End Phone Calls First.
Rule 7 Don't Accept a Saturday Night Date after Wednesday.
Rule 8 Fill Up Your Time before the Date.
Rule 9 How to Act on Dates 1, 2, and 3.
Rule 10 How to Act on Dates 4 through Commitment Time.
Rule 11 Always End the Date First.
Rule 12 Stop Dating Him if He Doesn't Buy You a Romantic Gift for Your Birthday or Valentine's Day
Rule 13 Don't See Him More than Once or Twice a Week.

Rule 14 No More than Casual Kissing on the First Date.

Rule 15 Don't Rush into Sex and Other Rules for Intimacy.

Rule 16 Don't Tell Him What to Do.

Rule 17 Let him take the Lead.

Rule 18 Don't Expect a Man to Change or Try to Change Him.

Rule 19 Don't Open Up Too Fast.

Rule 20 Be Honest but Mysterious.

Rule 21 Accentuate the Positive and Other Rules for Personal Ads.

Rule 22 Don't Live with a Man (or Leave Your Things in His Apartment).

Rule 23 Don't Date a Married Man.

Rule 24 Slowly Involve Him in Your Family and Other Rules for Women with Children

Rule 25 Practice, Practice, Practice! (Or, Getting Good at The Rules).

(Fein, 1995, p. 172).

The Rules does give some good advice instructing the female to boost their self-confidence and to act in a pro-active, assertive manner. At the same time, it also appears to inform her that this is isn't actually how she feels, simultaneously encouraging her to make sure she doesn't outwardly express these doubts.

> "You don't grovel. You're not desperate or anxious… you don't settle. You don't chase anyone…You're an optimist…Of course, this is not how you really feel. This is how you pretend to feel until it feels real. You act as if!" (Fein, 1995, p. 23).

Overall, the book appears to be about training males to enter an emotional state of being which Fein and Schneider call "longing." It does not, like the modern publications on mating, go into detail as to what lights up in brain activity when a male "longs" for a female. As a male, I could not quite come away from the book without the overall impression that the advice the authors were giving describes males as sexually hyper-active dogs that need to be trained to be patient before getting sex.

Now, don't laugh, because before his marriage to Emma Wedgewood, even the highly-educated, cultured and "honorable" gentleman Charles Darwin jotted down in an accountant's ledger the advantages and disadvantages of being married—here's some of what he penned as the positives:

"Constant companion, (& friend in old age) who will feel interested in one, — object to be beloved & played with. — Better than a dog anyhow—picture to yourself a nice soft wife on a sofa with good fire, & books & music perhaps" (van Wyhe, 2002).

We must remember that *The Rules* is not science, but cultural "wisdom" passed on from a small circle of individuals (one assumes all females) located in an unknown location at an unknown moment in time (starting in 1917?). These last two elements are extremely important, and as such, we really can't take the book seriously other than it may have reached a sizable amount of readers—the claim of "Over 1 Million Copies in Print" screams across the top of the book cover). What *The Rules* does not provide the reader is a balanced view of both the male and female's mating techniques, but it appears to provide the female reader with "sage wisdom" as to how to get the best "deal" in a relationship and marriage—and thus, the best benefits for her progeny.

Some critics will say that I have been a bit too harsh on the female authors discussing *The Rules* while males have their own, non-scientific "laddie" publications and media outlets that give advice in how woo the female and how to plant their seed via "scoring." But I have presented *The Rules* here because it presents a large chunk of "wisdom" that the female, at least in the view of the authors, takes a highly active role in the mating sequence, similarly to Perper's detailed conclusions—although, I view his work somewhat flawed because of its small sampling.

* * *

Obviously the emphasis in this chapter was to provide you with enough information to convince you that females have a far greater role in the mate selection process than Darwin's conclusion allowed, and that just as modern males need to believe this, so did he.

Things were simpler and more advantageous in Darwin's age for men, and they were highly restrictive for females. Society hemmed in the female as a reproductive entity with the major importance of producing a male heir, while making sure that she was forbidden to educate herself or own property; and as we know, Darwin's own personal objections to birth control were merely echoing the widely shared cultural beliefs of those within his social ranking and their religious beliefs. I will have more to say about Darwin and his objections to birth control in the conclusion.

In the end, there will be those who will take either the male or the female side of the mating sequence dance, but because of the tremendous risk and requirements of the reproductive responsibilities of raising children to maturity and giving those children the best advantages that life has to offer, there are no doubts, in my opinion, that the female has to be more highly selective in choosing a mate.

As a result, due to her physical evolution as "dainty" and the cultural emphasis on beauty, the female has undeniably evolved highly subtle body movements and tonal inflections that only the male she desires would hopefully understand and respond to. The evolutionary premise in today's modern world where propriety rules polite behavior, the female intuitively understands that if the male could pick up the subtlest of hints (such as a slight interest shown in facial expression), then he was intelligent enough to survive in a harsh world and therefore could be a suitable genetic contributor.

So who takes the lead in courtship? The male or the female?

I'm going to leave that final decision up to you, the reader because where you are sitting geographically on the planet in the present moment in time intuitively influences your behavioral options within the socially constructed norms for courting—be it in Los Angeles, Topeka, Chicago, Manhattan, London, Dubai, or Tokyo. You need to traverse this landscape and learn the "rules" of mating where you are located before seeking out that desirable mate.

But based on brain scans, educational opportunities, economic advances, and the freedoms available to females in our Western societies—including wide-spread availability of birth control—I would hazard a guess that the female here in the West, particularly in America, initiates the mating sequence 50 to 65% of the time. I am being cautious here, as some pundits believe that the female initiation exceeds 75%, and they staunchly believe the female absolutely controls whether the relationship proceeds or ends 100% of the time.

Darwin lived in a socially constructed world where the female had little freedoms in which to maneuver and determine her own future for herself and her children. And as you recall, in Darwin's historical timeline and at his hierarchical position, if a male did not take the lead in his family life by establishing a suitable home for the female and his children, along with assurances that they were safe and provided for, that male would face social ridicule and derision—pretty heavy stuff for a man to weigh. To those who believe that in Darwin's timeline it was "a man's world," just remember that this culture also had huge penalties for failure to achieve the ideal.

In the next chapter, I am going to introduce you to a female scientist whose work I also should have featured in Chapter 9 because of her recent work centered on the mating sequence and the brain, but quite frankly, it would have overloaded this chapter. And yet I have to let you know that when I began *Darwin's Flaw*, her work was so influential that there were many moments where I stopped and asked myself—what if?

CHAPTER 10

WHAT IF DARWIN WERE FEMALE?

Flipping history for insight

"...nothing is more important to a woman's future than the survival of her children... They are obliged to spread their DNA in perpetuity. This is nature's law"
(Fisher, 1995, p. 48 & 49).

As we near the end of our voyage, I want to introduce to you the work of a female scientist whose work I consider to be equal to Darwin in its evolutionary insight. She is alive and well and out there right now on the lecture circuit. No, she has not reached the level of celebrity that Darwin achieved in his day, and to the best of my knowledge she was not born into the financial and social hierarchical strata equal to that of Darwin's by marrying into a wealthy family. I am fairly certain that she does not have nine live-in servants equal to Darwin, and from all that I have been able to learn, her success is the direct result of her long hard work and studies.

Her name is Dr. Helen Fisher, and I have read most of her popular publications throughout the years. Her keen insight and intelligence, and of course, her female perspective—led me to believe that not only did the female lead us out of the jungle, but because of her innate abilities formed in our deep history, the female is on the cusp of regaining her leadership role for our entire species.

For the next few moments, let's use a bit of literary license and pretend that Dr. Helen Fisher was the equivalent of Charles Darwin in 1870. Let's pretend that her gender was the beneficiary of a long-held cultural understanding and appreciation of the female because of her reproductive abilities, while men—although once thought as a great benefit to humankind for their strength and protective nature—soon found disfavor amongst our species for their tendencies to form all-male alliances and use those collaborated strengths to commit wars against their fellow species. It was long known that of all the human mammals on Earth, only the male gender of chimpanzees and humans would conceive, organize, and implement the resource-depleting activity of war. These warring actions, of course, were done in the name of religion, racial purity, national security, or ambition. Let's pretend that our species continued to evolve onto a higher intelligence level and found effective means to end male-on-male aggression through the understanding the combined interactions of nature and nurture.

And let's also pretend that Dr. Fisher reached the same level of celebrity as Darwin. Let's pretend that she was born within the social hierarchies that nurtured and supported her gender from birth through death. Let's pretend that she lived in a home like Darwin's with nine live-in servants and was surrounded in an environment where only university educated females created all the inventions that powered the Industrial Revolution, using male strength as the axillaries to move heavy machinery and build skyscrapers.

Let's pretend that because of their aggressive nature, males were thought to be so low in intelligence that they were prohibited to attend schools of higher education, were forbidden to vote, inherit property, publish their thoughts, and were not allowed to take any leadership roles in society. This oppression of the male would be based primarily on long-held beliefs that the male was inferior to the female due to overwhelming evidence which showed that vast numbers of males engaged in anti-social behaviors and caused the prisons to be flooded with their numbers.

In this imaginary world we've created, the overriding cultural view in Dr. Fisher's timeline would believe that the enlarged brain of the male produced male tendencies towards violence, and that not all males were suited for domesticated protective roles within the family home; there were many calls that these males should be emasculated. Imagine that in some parts of the planet there were even calls that the male fetus be terminated in order to keep male populations to a minimum, thus creating an easily controlled demographic.

Let's pretend that King Albert—married to Victoria, Queen of England at that time—would broadcast to the nation: "Let the male be as a help-mate to the female, for nature has picked her to be the guiding light of our species. It is the male's role to be protector of future progenies of that guiding light."

In our fictional world, the female would be the head of the family not only because of her reproductive abilities but also because of their highly educated minds, and it would be the female who attended parliament and determined policies both foreign and domestic. Of course, we all know that males are superb chefs, so feeding the children in the home was never an adjustment. The difficult adaption for the male would be the childcare, which was once the sole provenance of the female. What kind of world would this be?

In this reversed gender role world, the male is about to take up the skills that the female has excelled at for millennia. Since we know that the human male has the largest penis of the three major primates, what if female heads of house kept her male not only as a protector and symbol of superior status but also for sexual pleasure? What might develop as a social construct concerning sexuality if the female gender was in total control? Would the females in this fictional world take the lead in mate selection and allow for multiple husbands with various roles? Would each male have a different role in the home? One for food preparation, one for child care, one for educating the children, and one for household chores inside and outside? And, of course, would there be a special male for solely for sexual pleasure? Or would some female-controlled households prefer sexual contact with all the males in her household so that when a child is born all would be involved emotionally?

* * *

Let's drop the fiction and see what Dr. Helen Fisher has revealed to us about the female in the 21st century. I need to praise Dr. Fisher, whose work has given my mind constant fuel in which to conceive that the female not only led us out of the jungle, but also, after traversing a bit of a bumpy path in history because of physical changes to her body and enduring generations of mental subjugations and being thought of as the 2nd sex, will once again take the leadership role for our species in the future.

* * *

All of the following quotes are all from Dr. Fisher's book, *The First Sex: The Natural Talents of Women and How They Are Changing the World*, 1995, Random House, New York:

"So here is my immodest proposal: As women continue to pour into the paid workforce in cultures around the world, they will apply their natural aptitudes in many sectors of society and dramatically influence twenty-first century business, sex, and family life" (p. xvi).

"From a strict evolutionary perspective this feminine proclivity to balance work and family makes sense: nothing is more important to a woman's future than the survival of her children...They are obliged to spread their DNA in perpetuity. This is nature's law" (p. 48 & 49).

"Women are built for mind reading. Touch, hearing, smell, taste and vision: all of women's senses are, in some respects, more finely tuned than those of men. Women also have a knack for decoding your emotions by looking at your face. They swiftly decipher your mood from your body posture and gestures. They remember more of the things in a room or office around you, putting you in social context" (p. 84).

"Women are, on average, more sensitive to touch... This feminine proclivity probably derives from millions of years of rearing babies...to detect its needs, the mother must touch her infant regularly. Cool? Rough? Supple? Rigid? Shaky? Soggy? For millennia, women needed sensitive fingertips to collect clues about their infant's health, thus selecting for women's sensitivity" (p.84-86).

"Women, on average, also have superior hearing… The answer lies in deep history. An ancestral woman needed superb hearing to rear her precious packet of DNA. Her child. Her baby's slightest whimper, its sigh, its troubled breathing: a woman had to distinguish among these inchoate sounds to know when her infant needed sleep or food or hugs" (p. 87A).

"Ancestral women probably used this superb sense to listen to their mates and lovers as well. They needed to distinguish if these men were honest, kind, and fatherly before they bore them babies and devoted their lives to rearing these men's genes" (p. 87B).

"Women's outstanding sense of smell probably evolved for the same evolutionary reason that women acquired their excellent sense of touch and hearing: to protect their young. With their impressionable noses, ancestral mothers detected dangerous smoke, rotting meat, even the scent of a stranger in the dark" (p. 88).

"Because of their ability to taste sweet, sour, salty, and bitter flavors in lower concentrations…This feminine acuity probably also evolved from women's primordial need to protect and nurture babies. With their ability to detect bitterness, ancestral mothers could guard against poison—since most poisons in the plant kingdom are bitter…With their ability to discern degrees of sweet and sour, mothers could reject unripe, less nutritious fruit. With their sensitivity to salt, our female forebears could recognize brackish water before they gave their child a drink" (p. 89).

"Women have superior eyesight…This is probably another legacy from deep history, when mothers needed good night vision to perform essential chores—feeding, doctoring, and comforting teary infants in the moonless grass" (p. 90).

"Women can also remember shades, tones, and color values more accurately than men. This color memory, as well as their superior ability to distinguish red and green, undoubtedly derives from women's long ancestry of foraging for fruit and vegetables…And as seasons turned into centuries, time selected for women with a superior color sense" (P. 91A).

"Women's ability to see shades of the color red many have helped them heal their infants, too…shades of red are also the color of fever, rash, and inflammation, of crying eyes and infected wounds. These ancestral females who could interpret the subtle hues of red in an infant's face and eyes detected early signs of sickness, as well as fear or shock" (p. 91B).

"Known as "location memory," this feminine ability to recall stationary items emerges with puberty, when estrogen levels rise…from bygone times when ancestral women were obliged to remember the location of water holes and the berry patches, termite mounds, and fig trees where they did their gathering" (p. 94).

"…women are the hands-on healers of everyday ills of the flesh. In fact, anthropologists have come to realize that in traditional societies mothers and grand-mothers are often the diagnosticians and healers, the primary providers of bodily well-being" (p. 113).

"Surely ancestral women...needed to coordinate emotionally with their young. Those who suffered when they saw a sick or unhappy infant devoted more time and energy to keeping this child alive... These children disproportionately lived—gradually selecting for women's superior ability to express sad-ness, pity, empathy, compassion, and other nurturing emotions" (p. 123).

"...women expend far more time at hands-on infant care than men do...And among almost all of our pri-mate relatives, females do all the nurturing of infants. To ensure that their young are cared for, women have evolved a powerful capacity to feel and express empa-thy" (p. 124).

These amazing clips from Dr. Fisher's writing are just the tip of the iceberg of her monumental work. How can anyone, female or male, not come away after reading these attributes and not believe that the female of our species was not only intelligent enough to lead our ancestors out of the deep history forest but would be better suited to lead our species in the future?

I want to conclude this chapter by recalling Darwin's 1870 posit about the male's superior intelligence and com-pare it with two paragraphs from Dr. Fisher:

Darwin wrote:

"To avoid enemies, or to attack them with success, to capture wild animals, and to invent and fashion weap-ons requires the aid of the higher mental faculties, namely, observation, reason, invention, or imagination. These various faculties will thus have been continually put to the test, and selected during manhood... *thus man has ultimately become superior to woman*" (Darwin, 1871, pp. 327-328, emphasis mine).

Dr. Fisher:

"Women have many exceptional faculties bred in deep history: a talent with words; a capacity to read postures, gestures, facial expressions, and other nonverbal cues; emotional sensitivity; empathy; excellent senses of touch, taste, smell, and hearing; patience; an ability to do and think several things simultaneously; a broad contextual view of any issue; a penchant for long-term planning; a gift for networking and negotiating; an impulse to nurture; and a preference for cooperating, reaching consensus, and leading via egalitarian teams" (Fisher, 1999, p. xvii).

"Ancestral women had the hardest job of any creature that trod the earth; raising long-dependent babies under highly dangerous conditions...Watch for snakes, listen for thunder, taste for poison, rock the sleepy, distract the cranky. Instruct the curious, soothe the fearful, inspire the tardy, and feed the hungry. Mothers had to do countless daily chores while they stoked the fire, cooked the food, and talked to friends" (Fisher, 1999, p. 12).

What conclusion would a fictional female Darwin posit in 1870?

"Thus woman has ultimately become superior to man."

What can we conclude from all of this?

CONCLUSION

WAITING FOR FEMALE
LEADERSHIP—AGAIN

As I present my closing arguments, I want to return to Darwin one last time by saying a few nice things about the old boy.

Darwin was a decent, kind, caring, and gentle male in regards to his family. He was meticulous in his observations regarding science and the rules surrounding publications and priorities. He shunned the glamour and limelight of celebrity after his return from his voyage on the Beagle— and especially after the publication of *Natural Selection*. He found it difficult to attend public meetings and rarely invited guests into his home because of his shy nature and, plausibly, the stomach ailments that plagued him throughout his life, which also seemed to flare up preceding public appearances if he attended them at all.

It has been argued widely that because of his passive personality, the stomach problems were psychosomatic and most likely aggravated by the vigorous cultural divide of beliefs between two powerful forces that dominated his timeline—science and religion. To complicate matters in his home life, his wife Emma, a devout Evangelical Christian, surely created some internal friction, but in the overall scheme of things, she supported his final decision to publish *Natural Selection* with the simple wifely advice of either publish or burn it.

If I have any criticism of Darwin, then it falls on the subject of birth control. And those two words are of vital importance as it feeds directly into *Darwin's Flaw*. Does the female have complete control over her body? Is the female free to make her own choice, or must the more "intelligent" male decide it?

In 1877, Charles Bradlaugh and Annie Besant penned a sixpenny pamphlet supporting birth control by using bleak Malthusian logic about "unrestrained gratification of the reproductive instinct" that would lead to over-populations, degeneration, and dissipation of the masses.

In April of that year, for distributing their "dirty filthy book," Bradlaugh and Besant were arrested (Chandrasekhar, 1981). Two months later, Bradlaugh wrote to Darwin asking for his help to testify on behalf of his and Ms. Besant's position thinking that since Darwin understood natural selection, he would logically uphold their opinion on the bleak prospects of humankind through overpopulation.

"Like Malthus, Darwin disparaged contraception, which he regarded as an impediment to natural processes. He thought easy access to contraception would lead to unfettered sexual activity outside marriage, which in turn would introduce licentiousness and vice, inadequate care of children, financial insecurity, death, and disease. "If it were universally known that the birth of children could be prevented, and this were not thought immoral by married persons, would there not be great danger of extreme profligacy amongst unmarried women?" (Browne, 2002, pg. 444).

Since we know that Darwin wrote over 15,000 letters during his lifetime, I decided that I would move toward concluding this first book of two with an open letter to the man responsible for giving birth to its inspiration.

AN OPEN LETTER TO CHARLES DARWIN:

Dear Charles,

I have a message for you 139 years after you expressed those negative views to George Arthur Gaskell, an advocate of birth control: birth control is now universally known and is being widely used by women all over the planet, and with it we are still functioning as a responsible, self-governing species. In fact, your insistence that birth control would lead unmarried women into "extreme profligacy"—shameless indifference to moral restraints—is a bit off the mark. There are no mass sightings of humans fornicating in the streets lured there by insatiable females.

In 2016, there is widespread, open availability of excellent and highly informative academic subjects—including sexual matters with explicit and non-prurient presentations. This is available via what we call the Internet (a kind of instant, electronic world-wide library composed of vast amounts of subjects that can be electronically retrieved right in front of your eyes and fingertips).

I am sure with your curious mind that you might feel overwhelmed by all the material that is available, and like a branch of evolution, you could get easily distracted from the path you needed to reach the top of the tree of life that you sketched out. You might risk the possibility that your ideas might perish on one of those far branches and get lost in the dust of history.

All this open information does contribute somewhat to what your Victorian society would consider "adverse moral behavior." But this open exchange of sexual material has vastly reduced or eliminated the shameful cloth that once draped all things sexual of your day. This deep, biological urge is nothing more than Mother Nature's way of remaining calm and carrying on with the larger task of the continuation of the species. It is one of the barriers placed before Mother Nature that individuals who populate highly-ranked social hierarchies create labels like "extreme profligacy" in a vain attempt to control populations considered "inferior" and "below" them in social hierarchies.

Who better to make the right decisions for our future progeny than the female, with whom each of us begins our voyage into this world? How can we not understand the female's intimate connection with Mother Nature and not respect her decision about her own body?

I hope history treats you well, Charles. You certainly have created quite a stir of emotions and behaviors in your path directly related to your life's work. But more importantly, you have contributed much to the debate of our species. I, for one, forgive you for this minor flaw of positing that the female makes no decision in selecting her future mate.

Be at peace.
Bill

* * *

I strongly believe what has dominated the gender and mating issues up to this point in our evolution has been the overwhelming adaptation of the female's need to provide food, safety, and a continuous and advantageous supply of resources for her progeny. I think the words of evolutionary psychologist Anne Campbell sums up the evolutionary pressures best as it relates to one female's evolutionary goals against other females:

"When push comes to shove and there is not enough to go around, I am afraid that it must be *my* progeny, *not yours* in the next generation" (Campbell, 2002).

So, looking back to the social dominance theory in Chapter 8, we can see much of this hierarchical discrimination against other populations considered "below" the higher-ranked female as an overall adaptation to provide the best resources for one's progeny over the "others," alongside the support of a male that agrees with the females overall perspective—or, is it the female that agrees with the male's overall perspective—or, both?

Looking over the horizon beyond the debate raised in this book about social hierarchies and the selection of male or female partners, there will be the larger, species-wide debate arguing which gender is best suited to control the wheel of our ship's rudder, leading our species into the vast ocean of future uncertainty called evolution.

Or will it be both genders?

And why is this important?

Because, as a species, we are faced with a larger issue—If we continue down this path, are we headed for mass extinction through global wars primarily started by one gender? I leave you with material for that debate from three evolutionary psychologists dating back to 1988 to stoke the present fire of debate over gender:

"When one restricts the focus to vertebrate species where *multi-individual* coalitions of males aggressively compete...only two species are known to exhibit warfare... common chimpanzees and humans (Tooby & Cosmides, 1988, pp. 2 & 3).

Or, to translate that academic posit so that we common people can understand it better, let's turn to the famous evolutionary psychologist, David Buss: "Men have recurrently engaged in warfare over recorded human history, whereas there is not a single documented case of women forming same-sex coalitions to go to war...Although more research is needed, the available empirical evidence supports the theory that men have evolved specific psychological mechanisms for warfare" (Buss, 2016).

So the final question that I ask of my readers would be: is there a gender that is better suited to guide our species into the future? Or should there be a new renaissance of thought that brings both genders toward a more equal footing concerning that future?

Obviously, I am in favor of correcting the scale of gender influence back towards a female-led worldview—at least for the next 250 years or so, when there may be another course adjustment.

Are you curious as to what a female-led future society would look like?

If so, I hope you will join me for my second book in the series: *Darwin's Flaw, 2.0: The Solution's Book,* where I will discuss how the female will guide us as she takes us on a course correction in the vast expanse called evolution.

APPENDIX

There is so much more material that could be added to this book. For example, supplementary to this Appendix, I have researched on YouTube and found some films that match or come close to the subjects in Darwin's Flaw. *To access this material, please visit www.DarwinsFlaw.com or simply snapshot this QR Code:*

REFERENCES

Ackerman, Jennifer, 2006—Bipedal-body http://ngm.nation-algeographic.com/print/2006/07/bipedal-body/acker-man-text

Allen, James, Hilton Als, Congressman John Lewis, & Leon F. Litwack, *Without Sanctuary: Lynching Photography in America*, 2000, Twin Palms Publishers.

BBC history—http://www.bbc.co.uk/history/historic_figures/darwin_charles.shtml

Biography.com, Darwin—http://www.biography.com/people/charles-darwin-9266433 A&E Television Networks, 235 East 45th St, New York, NY 10017.

Black, Ewin, *"The Horrifying American Roots of Nazi Eugenics"* September, 2003. http://historynewsnetwork.org/article/1796.

brainyquote.com, 1877A—Cecil Rhodes http://www.brainyquote.com/quotes/quotes/c/cecilrhode143204.html#XkL-0BlcbgoW3r2hh.99.

_____. 1877B—Cecil Rhodes http://www.brainyquote.com/quotes/quotes/c/cecilrhode175194.html.

Browne, Janet. *Voyaging: Volume I of a Biography*, 1995, Alfred A. Knopf, New York.

Browne, Janet. pp. 277 & 278. *Charles Darwin: The Power of Place*, 2002, New York, Alfred Knopf.

Buckley, Thomas & Alma Gottlieb. *Blood Magic: The Anthropology of Menstruation*. University of California Press, 1988.

Buss, David. *Evolutionary Psychology: The New Science of the Mind, 5th edition*. First published 2015, 2012, 2008 by Pearson Education, Inc. Published 2016 by Routledge, London and New York.

Cambridge University, 2013—via Hardman, Philippa, *Darwin's Women)*, www.youtube.com/watch?v=9qZxa3WjZQg 2013, Cambridge University.

Campbell, Anne. *A Mind of Her Own: the evolutionary psychology of women*, 2002, Oxford University Press.

Chandrasekhar, Sripati. *A Dirty filthy book*: the writings of Charles Knowlton and Annie Besant on reproductive physiology and birth control and an account of the Bradlaugh-Besant trial, 1981, University of California Press, Berkley.

Dahlber, Frances. *Woman The Gatherer*, quoting Dr. Adrienne Zihlman, 1981, Yale University Press, New Haven & London.

Dawkins, Richard, *The Selfish Gene*, 1976, Oxford University Press.

Darwin, Charles, *The Descent of Man, and Selection in Relation to Sex, 1871*. Modern Library, New York. Cited by Frances

Dahlber, p. 78. *Woman the Gatherer*, 1981, Yale University Press, New Haven & London.

_____. *The Descent of Man and Selection in Relation to Sex*. 1871, cited from the abridged version published by Dover Publications, 2010 with introduction copyright by Michael T. Ghiselin.

Delany, Janice, Mary Jane Lupton, and Emily Toth, *The Curse: A Cultural History of Menstruation,* 1976, Dutton, New York.

Dickens, Charles, *A Tale of Two Cities*, April 30 to November 29, 1859, a weekly series in All Year Round, published by Chapman & Hall, London.

Diamond, Jared, *Guns, Germs, and Steel: The Fates of Human Societies*, 1997, W.W Norton & Company, New York.

Ditext, 1563—Entire text of Statue of Artificers http://www.ditext.com/morris/1563.html

Elizabeththan.org, 2010. Statute of Artificers. http://elizabethan.org/compendium/80.html

English Heritage. *Down House: The home of Charles Darwin*, 1998, [reprinted 2000], Westerham Press Ltd.

Fein, Ellen and Sherrie Schneider, P. 1. *The Rules: Time-Tested Secrets for Capturing the Heart of Mr. Right*, 1995, Warner Books, New York, New York.

Fisher, Helen. *The Sex Contract: The Evolution of Human Behavior*, 1982, William Morrow & Company, New York.

_____. *Anatomy of Love: The Mysteries of Mating, Marriage, and Why We Stray,* 1992. A Fawcett Columbine book, Published by Ballantine Books, New York.

_____. *The First Sex: The Natural Talents of Women and How They Are Changing the World,* 1999, Random House, New York.

Fletcher, Anthony. *Gender, Sex & Subordination in England, 1500-1800,* 1995, Yale University Press, New Haven and London.

George, Rose, December 28, 2012, "The Taboo of Menstruation," Rose George, *The New York Times.*

Givens, David, pp. XVI & XVII *Love Signals: A Practical Field Guide to the Body Language of Courtship, 2005,* St. Martin's Press.

Goodall, Jane, 2010, The Goodall Institute—Termite Fishing. https://www.youtube.com/watch?v=inFkERO30oM

Houppert, Karen, *The Curse: Confronting the Last Unmentionable Taboo: Menstruation,* 1999, Farrar, Straus and Giroux, New York.

Hrdy, Sarah Blaffer. *Mothers and Others: The Evolutionary Origins of Mutual Understanding,* 2009, The Belknap Press.

Hudson, Valerie M. and Andrea M. den Boer, *Bare Branches: The Security Implication of Asia's Surplus Male Population,* 2004, MIT Press, Cambridge, Massachusetts.

Huxley, Thomas, p. 197. *On the Reception of the 'Origin of Species: The Life and Letters of Charles Darwin, Vol. 2,* 1887, including an autobiographical chapter, John Murray, London.

Knight, Chris, p. 394. *Blood Relations: Menstruation and the Origins of Culture,* Yale University Press, 1991. Citing Kirahara, M. 1982. *Menstrual Taboos and the importance of Hunting.* American Anthropologist 84: 901-3.

Malick, Amy, August, 29, 2015. http://abcnews.go.com/Health/story?id=116833

Mascia-Lees, Frances E. and Nancy Johnson Black. *Gender and Anthropology,* 2000, Waveland Press, Inc. Here, Mascia-Lees and Johnson provide the following references: [Buton1972]—Buron, F. 1972. "Sexual Climax in Female *Macca mulatta." Proceedings of the Third World International Congress of Primatology* 3:180-91; [Hrdy 1979] Hrdy, Sarah B. 1979. "The Evolution of Human Sexuality: The Latest Word and the Last." *Quarterly Review of Biology,* 54:309-14; [Manson 1986], Mason, W. C. 1986. "Sexual Cyclicity and Concealed Ovulation." *Journal of Human Evolution* 15:21-30; and [Rowell 1972], Rowell, T. 1972. "Female Reproductive Cycles and Social Behavior in Primates." *Advances in the Study of Behavior* 4:69-105.

Miller, Geoffrey. *The Mating Mind: How Sexual Choice Shaped the Evolution of Human Nature,* 2000, Doubleday, New York.

NYU. http://www.nyu.edu/library/bobst/research/fales/exhibits/wilde/6club.htm.

Perper, Timothy. *Sex Signals: The Biology of Love*, 1985, iSi Press, Philadelphia, 1985.

Remmel, Ethan, May - June 2008—Benefits of a long childhood. http://www.americanscientist.org/bookshelf/pub/the-benefits-of-a-long-childhood

Rosen, 2010A- Posted by Dr. Bruce Rosen, September 30, 2010 at http://vichist.blogspot.com/2010/09/west-end-club.html

Rosen, 2010B—Posted by Dr. Bruce Rosen, September 01, 2010 at http://vichist.blogspot.com/2010_09_01_archive.html

Sidanius, Jim & Felicia Pratto. *Social Dominance: An Intergroup Theory of Social Hierarchy and Oppression*, Cambridge University Press, 1999. Cambridge, United Kingdom.

Sagan, Carl. *Cosmos: A Personal Voyage*. Random House, 1980.

Spencer, Herbert. *Principles of Biology, Vol I, 1864*, William & Norgate. London.

Steinem, Gloria, essay, "If Men Could Menstruate," *Ms. Magazine*, October, 1978.

Tate, Karen. Scared Places of Goddess: 108 Destinations, 2006, Consortium of Collective Consciousness.

Tooby, John & Leda Cosmides, *The Evolution of War and its Cognitive Foundations,* pg. 2 & 3, Institute for Evolutionary Studies Technical Report 88-1, 1988.

Tosh, John. *A Man's Place: Masculinity and the Middle-Class Home in Victorian England,* 1999, Yale University Press, New Haven.

Tyson, Neil DeGrasse, 2014, The Unofficial Guide to Cosmos: Fact and Fiction in Neil DeGrass Tyson's Landmark Science Series, Discovery Institute.

van Wyhe, ed. 2002 'This is the Question—Marry Not Marry' Memorandum on marriage. http://darwin-online.org.uk/

Wedekind, Claus, et al. 260 (1359): 245-249). *"MHC-dependent preferences in humans."* 1995, Proceedings of the Royal Society of London.

Wiesner-Hanks, Merry E. p.17, *Gender in History: Global Perspectives,* 2nd Edition, 2011, Wiley-Blackwell. A John Wiley & Sons, Ltd publication, Chichester, West Sussex, United Kingdom.

Zahavi, A., 1975—http://www.frozenevolution.com/handicap-hypothesis-evolution-female-preferences/

Many people are not aware that Darwin's natural selection theory was co-authored and presented to the world in 1858 by what Darwin's biographer, Janet Browne, called, "a nobody from nowhere." His name was Alfred Russel Wallace.

Mr. Spriggs considers himself in the same category as Mr. Wallace in presenting an overarching theory that seems to have been overlooked by many credentialed scientists and that may astonish most readers: that the female of our species led us out of the jungle; that she had to adapt to many physical changes to her body; that she endured insults claiming her to be inferior to men; that she was denied access to education, property, and publishing her own thoughts; that overall she existed in a totally male-dominated world while caring, guiding, and advancing her progeny.

"I've held a long-term view bordering on astonishment concerning the incredible innate abilities of the female and her ability to adapt and survive in a highly restrictive world while at same time attempting to give her progeny all the best that she can provide. This book is about male-kind's blind eye to those innate abilities and our species' bright future with the female at the helm—again."

But to those who criticize this theory and want to limit or destroy its validity—relax—Mr. Spriggs leaves us with a timetable to suggest that this won't happen for another 200 to 250 years.

Mr. Spriggs continues his work on the societal changes on the horizon for our planet with work on his second book in the series: *Darwin's Flaw, 2.0: The Solutions Book.*

www.ingramcontent.com/pod-product-compliance
Lightning Source LLC
Chambersburg PA
CBHW030940180526
45163CB00002B/645